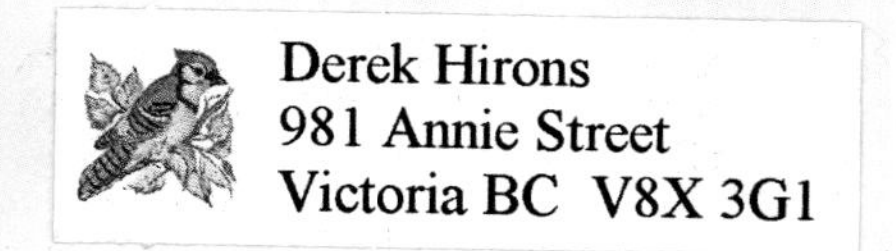

SPACE EXPLORATION 2007

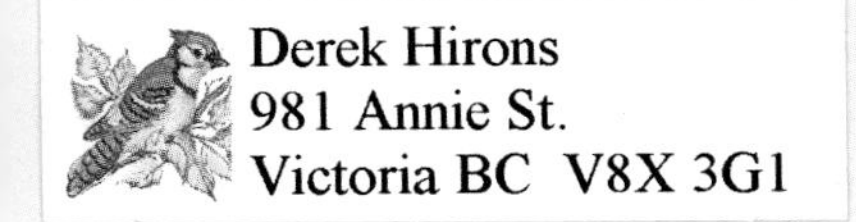

Brian Harvey

SPACE EXPLORATION 2007

Brian Harvey, M.A., H.D.E., F.B.I.S.
2 Rathdown Crescent
Terenure
Dublin 6W
Ireland

Front cover illustration: The first ever landing by a spacecraft on the moon of another planet! An artist's impression depicting the parachute descent of the European Space Agency's Huygens lander through the cloudy atmosphere of Saturn's largest moon Titan to its eventual historic touchdown on the surface. Image courtesy C. Carreau and the European Space Agency.

SPRINGER-PRAXIS BOOKS IN SPACE EXPLORATION
SUBJECT ADVISORY EDITOR: John Mason B.Sc., M.Sc., Ph.D.

ISBN 10: 0-387-33330-4 Springer Berlin Heidelberg New York
ISBN 13: 978-0-38733330-4 Springer Berlin Heidelberg New York

Springer is a part of Springer Science + Business Media (springeronline.com)

Library of Congress Control Number: 2006928110

Cover design and cartoons: Jim Wilkie
Typesetting and design: BookEns Ltd, Royston, Herts., UK

Printed in Germany on acid-free paper

CONTENTS

NASA

Preface

WELCOME TO SPACE EXPLORATION, 2007

WELCOME TO *Space Exploration 2007*, the first of a series of annuals bringing readers the latest news, views, information and comment on space exploration. This book is the first in an annual series published by Praxis/Springer and aims to provide a balanced content of news and articles covering all the countries involved in space exploration, from the huge programme run by the United States to relative newcomers like India. The annual will look back to the great events in past space exploration, to future planned missions, such as the Bush plan to return to the moon and fly onward to Mars. The book will report on both manned, piloted missions and the pioneering voyages of unmanned space probes. The annual will cover the unspectacular but nonetheless important area of the application of space technology and how this can improve the quality of life on Earth.

For the first *Space Exploration* annual, a team of writers has presented a series of chapters on the cutting edge of manned and unmanned ventures into space today. First, many readers will have seen the International Space Station cross the night sky and Neville Kidger describes the progress in building the space station so far, with the promise of what's still to come. Second, we turn to the solar system and the remarkable progress made in its exploration over the past number of years. In chapter 3, David Harland writes of the results from the recent explorers of Mars. Chapter 4 is called *Solar system log* and Sean Solomon and Ralph McNutt, who are both leaders of the MESSENGER project, describe the United States' first probe to the Mercury in over thirty years. David Harland brings us results of exploration from two different ends of the solar system: Venus and Saturn. Rosaly Lopes writes about results from Titan while John Mason describes the extraordinary missions made to comets and asteroids over the past two years. Rick Greenberg focuses on the exploration of Jupiter's icy moons.

Chapter 5 looks forward to future space missions and especially the Vision for Space Exploration. John Catchpole describes how the American space agency, NASA, plans to return to the moon and fly astronauts to Mars, while Jim Oberg, in a thought-provoking article, looks at another, possibly better way of developing manned space exploration outside Earth orbit. The next chapters also look at future developments. Laurent de Angelis describes the new Soyuz launchpad in the south American jungle while Paolo Ulivi previews the forthcoming moon missions by China and India. Finally, it is easy to forget that Russia remains, in terms of launches made every year, the leading spacefaring nation in the world. Here, in chapter 8, Bart Hendricx examines Russia's plans for a new space shuttle, the Kliper.

Thanks are due to the writers who contributed, as well as the space agencies and individuals who provided illlustrations, such as NASA, ESA (www.esa.int), JAXA and Nicolas Pillet.

But first, in chapter 1 a review of important developments in 2005.

BUNNY
IN SPACE
ASTRO KAT
ROCKET MANUAL
CAT FOOD

PUBLISHER'S NOTE

IN MY youth I was an avid collector of 'annuals' which, in the UK, consisted of a large-format book containing strip cartoons, short stories and other items and puzzles designed to stimulate the imagination and therefore the learning process.

In developing the range of books in the Praxis imprint *Space Exploration* it occurred to me there was a real need for a book for enthusiasts of all ages on space exploration. I found a kindred spirit for this idea in Brian Harvey who is also a lover of annuals.

This provided the ideal opportunity to fulfill a longstanding ambition of mine to produce a book - an annual - containing various reviews of latest developments of discoveries in our Solar System. And let's illustrate the chapters with a series of one-page cartoons on the various topics, to lighten the learning curve and make the reader smile.

The cartoons show how the latest space missions, anniversaries and the people behind the missions could be seen through the eyes of a cat! This allowed me to have some 'publishing fun' with our cover designer and illustrator, Jim Wilkie, in developing this idea and thus immortalising my wife Jo's incredible 20-year old Russian Blue cat, Bunny, as AstroKat!

Sincere thanks go to Jim's wife, Rachael, for helping to shape and focus the cartoon ideas with Jim and to Arthur Foulser of BookEns, who surpassed the design challenge I set him for the layout of the text. To our intrepid *Space Exploration* Advisory Editor, John Mason, a big thank you for his work on the final selection of images and other essential detail.

I've always believed learning should be fun, so for the first time since I started the Praxis imprint over a decade ago, I decided to write a publisher's note to explain the thinking behind this book. To all readers, please enjoy this annual and other volumes in future years. This is the start of a classic series and the very first volume usually becomes a prime collector's item, so you should keep your copy as a treasure after reading it.

Finally, a big thank you to Harry Blom of Springer New York for his enthusiastic support for this project.

Clive J Horwood

The Space Shuttle Discovery and its seven-member crew launched at 10:39 a.m. (EDT) on 26 July 2005 to begin the two-day journey to the International Space Station on the historic Return to Flight STS-114 mission. Image courtesy NASA.

1

THE EPIC JOURNEY BEGINS...

The space age began in October 1957, and every year brings its own crop of exciting developments, new records, trends, launches and anniversaries.

Here *Brian Harvey* sets the scene with a review of the major events during 2005.

The Annual REVIEW

One of the dates that separate the old world from the new is 1957. On 4 October 1957, fifty years ago, the Soviet Union launched the first spacecraft, Sputnik. Early visionaries, going back to Herman Oberth and science fiction writers such as Jules Verne had dreamt of how people might one day fly in space. The first modern rocket, Wernher von Braun's A-4, known and feared as the V-2, was flown over the Baltic on 3 October 1942. Once the war was over, European, American and Russian engineers and scientists realized that the A-4 had brought technology to a level which could contemplate the putting into orbit of a small satellite. The early 1950s saw leading American companies, designers and think tanks put forward proposals for a small satellite that could orbit the Earth. In the Soviet Union, a satellite team was assembled by one of Russia's greatest designers, Mikhail Tikhonravov. They tried to interest their respective governments. But the cold war dominated the thinking of the two superpowers and their rulers were principally interested in missiles, not scientific satellites of questionable value.

The International Geophysical Year (IGY) provided a practical opportunity and political environment that would make the first satellite possible. Setting aside political differences, governments of the world declared 1958 a year for exploring the Earth's geophysical environment and the participant countries announced a series of experiments and observations based on land, sea and polar ice, using different kinds of equipment for observations and even using small sounding rockets. The promoters of scientific satellites eventually persuaded their governments that a small set of instruments placed in Earth orbit would be the perfect way to mark the geophysical year. An announcement by the United States government that it would launch a satellite during the year led to a similar announcement in Moscow and what became a race got under way.

The United States were the obvious winners of this race. The United States were much wealthier and had suffered none of the appalling damage to their homelands as had the Soviet Union during 1941-5. The American economy expanded throughout the 1950s, funding a boom in home-building, road construction and consumer spending.

Above: *The team that built the first Sputnik 50 years ago.*

America's scientific dominance was self-evident. The ranks of American scientists had not been decimated by purges. The Americans had taken the most knowledgeable of Germany's scientists westward, with their formidable experience not just in rocket engines but the associated disciplines of guidance, control and precision engineering. When the federal government eventually approved the satellite project, it opted for a civilian satellite put up by new rocket developed by the United States Navy, Vanguard, in preference to a development of the A-4 offered by von Braun. The decision was a complex one, shaped by inter-service rivalry, military vs civilian satellites and personal preferences.

A series of announcements by the Soviet Union of its plans to launch a satellite to mark the geophysical year attracted little attention - either in the west or the USSR itself. Western nations probably doubted that the Soviet Union had the capacity to do such a thing, neglecting recent demonstrations of scientific prowess such as its mastery of atomic power and the Tupolev 104 jetliner. Within the USSR, people had been brought up for years on the ideas of scientists and theorists. There had been science fiction films, space exhibitions and clubs for space travel since the 1920s.

Originally, the USSR planned to put up a large scientific satellite which the designers called 'object D'. It weighed 1.5 tonnes, marking a dramatic contrast to Vanguard's 1.5kg. The design of object D proved so difficult that chief designer Sergei Korolev substituted a much smaller, simpler satellite, the PS or Preliminary Satellite. He asked the builders to equip it with aerials so that its transmitter could be picked up by simple receivers over long distances. 'Object D' was not the only cause of difficulty on the Soviet side, for the rocket to launch it, the R-7, blew up several times during tests, not eventually flying until August 1957, its nosecone successfully impacting in the Pacific ocean.

Sergei Korolev and his colleagues left us their memories of what happened on that never-to-be-forgotten day. They tell us of how Sputnik was placed on the top of its R-7 rocket, lying flat on its railcar in the steel hangar in Baikonour cosmodrome, Tyuratam. For the last time on Earth, the plug was taken out of the Sputnik to test its transmitter and the *beep! beep! beep!* could be heard, on amplifier, echoing around the hangar. The plug was reconnected and it fell silent. The next time the plug would be disconnected was to be in orbit and set the transmitter beeping again.

The first Earth satellite took off into a cold night sky. It was nearly midnight and Korolev and his colleagues watched the R-7's flames disappear into the high distance over the Kazakhstan desert to the east. There was nothing they could do now but wait for the 90mins that it took for Sputnik to circle the Earth and return over the launch site. After what must have been a lifetime, they gathered in the hangar to listen.

And eventually, they came, the tiny *beep! beep! beep*! sounds through the dark distance. Their joy knew no bounds and Korolev climbed a step ladder to make a short speech to his ecstatic colleagues. For many of them, getting the Sputnik into orbit was the culmination of their life's work. What Korolev told them was probably not what they expected, for he told them that this was just the beginning of a new life's work that would send cosmonauts into orbit and on to Mars.

Was this the excited reaction of the Kremlin? Not a bit of it. Korolev told Khruschev who told a party meeting of the accomplishment and then went to bed. The following morning's *Pravda* put the launching well down the front page, with the

Right: *Chief designer Sergei Korolev.*

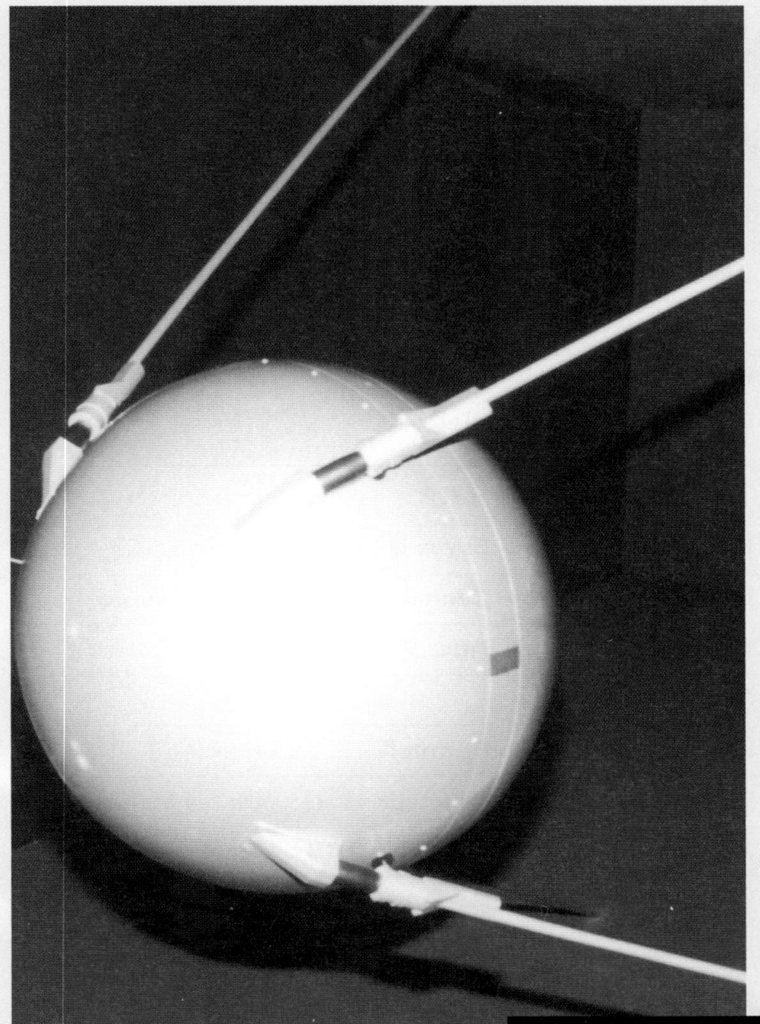

page. Across the world, the transmitter's little beep could be picked up by amateurs with little difficulty. With unintentional but exquisite timing, the launching came just as the world's space scientists were assembling for their annual conference in Barcelona, Spain. In the United States, phone calls rang during all day and all night as politicians, administrators and generals called one another to relay the stunning news. The political panic quickly set in. Children and adults went out into their back yards to watch the Sputnik (or more likely, its larger rocket stage) a pinpoint of light silently cross the dark autumn skies of the northern hemisphere.

There were only three Sputniks, a Russian word that means 'companion'. Sputnik 2, a similar design but with an animal cabin, carried the dog Laika into orbit a month later, the first living creature to fly (and sadly, to die) in space. Object D was the scientific laboratory planned as the original Sputnik and flew in May 1958 as Sputnik 3. But nothing was ever the same again after the first Sputnik on the night of 4 October 1957.

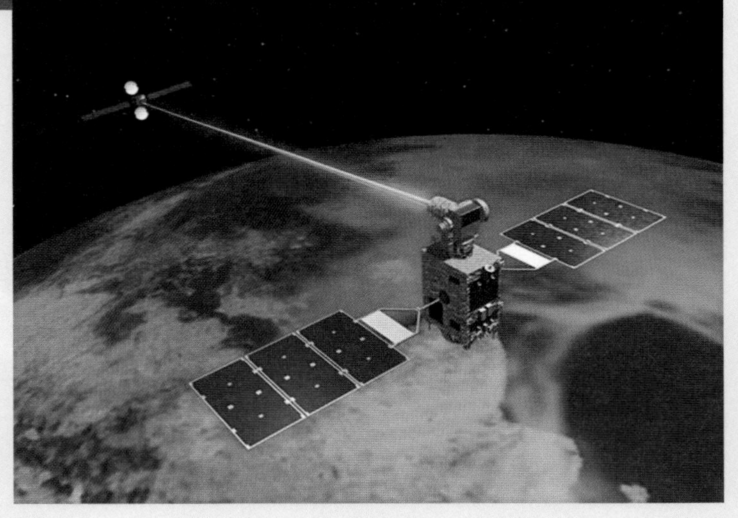

bland heading of *Tass communiqué*, followed by three paragraphs. The British Broadcasting Corporation announced the launching at the end of the midnight news, but the tone of the announcer was uncertain, as if not knowing what to make of the event or its significance. Korolev and his colleagues prepared to take the long train journey back to Moscow.

It was only once they were on their way that the significance of the launching began to hit home. As the train chugged across the Kazakh steppe and the lowlands of the Volga, crowds surged forward to congratulate the designers on their achievement, their numbers and enthusiasm growing all the time. It took the leaders of the Kremlin less than a day to sense the pride which ordinary Soviet people took in the Sputnik. On 6 October 1957, the satellite was the only story on the *Pravda* front

Top: *Sputnik I.*

Above: *Japan's OICETS experimental communications satellite. Image courtesy JAXA.*

LAUNCH LOG 2005

Date	Payload	Launch site	Launch vehicle	Country
12 Jan	Deep Impact	Cape Canaveral	**Delta II**	*United States*
20 Jan	Cosmos 2414 Tatania	Plesetsk	**Cosmos 3M**	*Russia*
3 Feb	AMC-12	Baikonour	**Proton M**	*Russia*
3 Feb	USA 181	Cape Canaveral	**Atlas IIIB**	*United States*
12 Feb	XTAR Maqsat B2 Sloshsat	Kourou	**Ariane 5**	*Europe*
26 Feb	MTS-1	Tanegashima	**H-IIA**	*Japan*
28 Feb	Progress M-52	Baikonour	**Soyuz U**	*Russia*
1 Mar	XM-3 Rhythm	Odyssey Platform	**Zenit 3SL**	*Russia/Ukraine*
11 Mar	Inmarsat 4	Cape Canaveral	**Atlas V**	*United States*
28 Mar	TEKh 42	ISS		*Russia*
30 Mar	Express AM-2	Baikonour	**Proton K**	*Russia*
11 Apr	XSS	Vandenberg AFB	**Minotaur**	*United States*
12 Apr	APstar 6	Xi Chang	**Long March 3B**	*China*
15 Apr	DART	Vandenberg AFB	**Pegasus**	*United States*
15 Apr	Soyuz TMA-6	Baikonour	**Soyuz FG**	*Russia*
26 Apr	Spaceway F-1	Odyssey Platform	**Zenit 3SL**	*Russia/Ukraine*
29 Apr	Lacrosse/Onyx	Cape Canaveral	**Delta IV**	*United States*
30 Apr	USA 182	Cape Canaveral	**Titan IVB**	*United States*
5 May	Cartosat Hamsat	Sriharikota	**PSLV**	*India*
20 May	NOAA-18	Vandenberg	**Delta II**	*United States*
22 May	DirectTV-8	Baikonour	**Proton M**	*Russia*
31 May	Foton M-2	Baikonour	**Soyuz U**	*Russia*
16 June	Progress M-53	Baikonour	**Soyuz U**	*Russia*
24 June	Intelsat Americas/Telstar 8	Odyssey Platform	**Zenit 3SL**	*Russia/Ukraine*
24 June	Express AM3	Baikonour	**Proton K**	*Russia*
6 July	Shi Jian 7	Jiuquan	**Long March 2D2**	*China*
10 July	Astro E-2	Kagoshima	**M-V**	*Japan*
26 Jul	Discovery	Cape Canaveral	**Shuttle STS-114**	*United States*

2 Aug	**FSW 21**	Jiuquan	**CZ - 2C**	*China*
11 Aug	**Thaicom4**	Kourou	**Ariane 5**	*Europe*
12 Aug	**MRO**	Cape Canaveral	**Atlas V**	*United States*
13 Aug	**Galaxy 14**	Baikonour	**Soyuz FG**	*Russia*
24 Aug	**OICETS** **INDEX**	Baikonour	**Dnepr**	*Russia*
26 Aug	**Monitor E**	Plesetsk	**Rockot**	*Russia*
29 Aug	**FSW 22**	Jiuquan	**CZ - 2D2**	*China*
2 Sep	**Cosmos 2415**	Baikonour	**Soyuz U**	*Russia*
8 Sep	**Progress M-54**	Baikonour	**Soyuz U**	*Russia*
9 Sep	**F-1R**	Baikonour	**Proton M**	*Russia*
22 Sep	**Streak/USA 185**	Vandenberg	**Minotaur**	*United States*
26 Sep	**GPS-IIIR**	Cape Canaveral	**Delta II**	*United States*
30 Sep	**Soyuz TMA-7**	Baikonour	**Soyuz FG**	*Russia*
12 Oct	**Shenzhou 6**	Jiuquan	**CZ-2F**	*China*
13 Oct	**Syracuse 3a** **Galaxy 15**	Kourou	**Ariane 5**	*Europe*
19 Oct	**USA 183**	Vandenberg	**Titan IVB**	*United States*
27 Oct	**Mozhayets 5** **Sina 1** **China DMC 4** **SSET Express: XIVI,** **UWE, NCube 2** **Topsat**	Plesetsk	**Cosmos 3M**	*Russia*
8 Nov	**Inmarsat**	Odyssey Platform	**Zenit 3SL**	*Russia/Ukraine*
9 Nov	**Venus Express**	Baikonour	**Soyuz FG Fregat**	*Russia*
16 Nov	**Spaceway 2** **Telkom 2**	Kourou	**Ariane 5**	*Europe*
21 Dec	**INSAT 4A** **MSG-2**	Kourou	**Ariane 5**	*Europe*
21 Dec	**Progress M-55**	Baikonour	**Soyuz U**	*Russia*
21 Dec	**Gonetz D1M** **Cosmos 2416**	Plesetsk	**Cosmos 3M**	*Russia*
25 Dec	**Cosmos 2417-9**	Baikonour	**Proton K**	*Russia*
28 Dec	**Giove 1**	Baikonour	**Soyuz FG Fregat**	*Russia*
29 Dec	**AMC-23**	Baikonour	**Proton M**	*Russia*

Here are some details. Starting with the first launch of the year, Cosmos 2414 was a military navigation satellite in the *Parus* series. Tatiana was a 23kg microsatellite to celebrate 250 years of the Moscow Lomonosov State University and designed to study magnetic fields (the satellite is also called Universiteski).

USA 181 was a classified American satellite operated by the National Reconnaissance Office and used for maritime surveillance. A subsatellite may have been deployed early in the mission. This was the last launch of the Atlas IIIB.

MTS-1 was Multifunctional Transport Satellite and marked the return to flight of Japan's H-IIA rocket. The H-IIA is Japan's main large launcher and had been grounded after a failure in November 2003. A vigorous programme of quality control had been put in place since, with an evidently successful outcome. MTS-1 is a satellite that combines functions of weather forecasting and air traffic control.

DART, or Demonstration of Automated Rendezvous Technology was a small 140kg experimental satellite intended to spend 18 months intercepting satellites and rocket stages at altitudes of up to 850km. There is nothing new about rendezvous in space: Vostok 3 and 4 passed close to one another in 1962. The Americans achieved the first rendezvous in 1965 (Gemini 6 and 7) while Russia's Cosmos 186 and 188 rendezvoused and docked together in 1967. The principal purpose of DART was to test out the capabilities of autonomous computer systems. All did not go well though, for although **DART** closed to within 100m of an old satellite launched in 1999, it soon became clear that it had used up nearly all of its fuel in doing so. NASA termed the mission a partial success - and a good learning experience for when such manoeuvres must be practised in Martian orbit in the future. Sent up only four days before DART, XSS was a small military satellite to test out rendezvous, autonomous and inspection technology.

Above: *Giove-1, the first satellite in Europe's new satellite navigation system was launched by a Soyuz Fregat launcher in December 2005. Image courtesy ESA*

USA-182 was a Lacrosse Onyx classified American radar imaging satellite.

Cartosat was a 1.5 tonne Earth resources and mapping satellite while Hamsat is a small, 42.5kg amateur radio satellite.

NOAA-18 was a weather satellite for the National Oceanic and Atmospheric Administration (NOAA) after which it is named and will collect data on the Earth's atmosphere, seasonal development and climate change.

Astro E-2 was Japan's fifth astronomical observatory in this series, replacing the earlier

Astro E-1 which failed at launch in February 2000. In line with the Japanese tradition of renaming satellites once they reach orbit, Astro E-2 was renamed Suzaki after a bird god. Suzaki is a 1,700kg observatory circling at 550km carrying an x-ray spectrometer, x-ray imaging spectrometer, hard-x-ray detector and five x-ray telescopes.

MRO was Mars Reconnaissance Orbiter (see David Harland: *Arrival at the red planet*, Chapter 3). It was the first of two interplanetary missions, the other being Venus Express, launched by Russia in November.

OICETS (Optical Inter Orbit Communications Engineering Test Satellite) was a Japanese satellite used to test laser communications systems in orbit. Originally intended for the cancelled Japanese J-1 launcher, OICETS was eventually transferred to a Russian Dnepr. Once in orbit it acquired the name Kirari. Accompanying OICETS was INDEX, the Innovative Technology Demonstration Experimental Satellite, which was given the name Rimei.

Cosmos 2415 was a recoverable military observation satellite in the *Kometa* series introduced in 1981 and used to compile high-accuracy military topographical maps. It was the first *Kometa* for five years.

Monitor was a new type of Earth observation satellite. Traditionally, Russian observation satellites were large and with Monitor, Russia follows Europe and other countries in flying smaller, high-quality observation systems on less expensive rockets.

Streak was a technology demonstrator with instruments to measure atomic hydrogen and atmospheric density.

GPS IIIR was a modernized version of the American Global Positioning System navigation satellite.

Syracuse 3a was a French military communications satellite.

The October Cosmos 3M launch from Plestsk had a complex set of payloads. The prime payload, the 80kg Mozhayets 5 was a training satellite built by the students of St Petersburg military academy, but it failed to detach from the rocket carrier and no signals were ever received. Topsat was a 120kg demonstrator imaging satellite of 2.5m resolution built by SSTL in Surrey, England, for the British Ministry of Defence. DMC-4 was a 150kg small satellite built by SSTL as China's part of the Disaster Monitoring Constellation and carries two cameras. SSETI Express was a European technology demonstrator built by students from 23 universities, but it suffered an electrical failure after a day in orbit. This carried three picosatellites, each weighing in at 1kg! NCube was built by Norwegian students while UWE was built by the University of Wurzberg to test telemetry, sensors and guidance. Cubesat XI-V was built by the University of Tokyo's Intelligent Space Systems Laboratory. The precise nature of the Sina-1 payload is uncertain. It was built by the OKB Polyot company in Omsk for Earth observations, but it has also been identified as an Iranian satellite.

USA 183 was a 19-tonne polar orbiting KH-11 digital photo-reconnaissance satellite, the fourteenth in the series begun in 1976 and which looks like the Hubble Space Telescope in appearance. The KH-11 series replaced the KH-9 series, which sent down recoverable film capsules in pods. USA 183 was expected to go into an orbit of 264 - 1,050km, 98°. This was the last launch of a Titan IVB rocket.

The following satellites were commercial launches by Russia: Galaxy 14, Anik F1R (for Canada) and AMC-23. The Zenit 3SL is used entirely for commercial satellite launches: *Rhythm* digital radio (following predecessors XM-1, *Rock* and XM-2, *Roll)*, Spaceway F1 high-definition TV and Inmarsat 4. The United States launched Inmarsat 4. The European Space Agency launched Thaicom 4, Galaxy 15, Spaceway 2, Telkom 2.

The following were domestic communications satellite launches: APstar 6 (China) and Express AM2 (Russia).

The last European launching of 2005 put two applications satellites into orbit: INSAT 4A, which will supply direct television throughout India and the second Meteosat Second Generation, a European weather satellite.

Gonetz D-1M was a Russian military communications satellite. Cosmos 2416 was a small military satellite. Cosmos 2417-9 were three satellites in the Russian navigation satellite system, GLONASS. These launches brought the strength of the GLONASS constellation to 17, two thirds of its intended strength of 24 satellites.

Giove 1 was the first satellite in Europe's new navigation satellite system called Galileo. Giove 1 was a test satellite built by SSTL in Surrey, England.

In addition, the TEKh nanosatellite, weighing only 4.5kg, was launched by the crew of the International Space Station from orbit in the course of a spacewalk on 28 March.

Highlights of 2005

January: *a year on Mars*

January 2005 marked a full year since NASA's two rovers, Spirit and Opportunity, touched down on Mars. It was an anniversary no one expected, for the rovers had been designed to last only 90 sols (a sol is a Martian day, 24hr 37mins, slightly longer than the Earth day). These remarkable rovers were not only roving a year later, but had survived all kinds of major and minor problems, from misbehaving computers to getting stuck in sand dunes. In its first year to January 2005, Spirit travelled 4km and Opportunity 2km. Spirit, which landed in crater Gusev, climbed up the Columbia Hills (named for the crew members of the Space Shuttle *Columbia* who died on 1 February 2003) to send us back stunning images of the 65km wide crater floor. Opportunity, which landed in Meridiani, explored craters and rocks. Opportunity went on to set a record of 220m covered in a single day (see David Harland: *Arrival at the red planet*, Chapter 3).

February: *Europe's comeback*

On 12 February, the European Space Agency introduced the new version of its Ariane 5 rocket, which first flew successfully in 1997. This was called the ECA model and the successful launching was a great relief, for the first ECA mission had failed in December 2002. The ECA model introduced the Vulcain 2 engine, with 20% more thrust than Vulcain 1; an extra 60 tonnes thrust on its solid rocket engines; and a new cryogenic upper stages, the HM-7B. The mission put into orbit the XTAR Spanish military communications satellite; an instrumentation model called Maqsat B2; and the Sloshsat experimental satellite to test the behaviour of liquids in orbit.

Above: *Europe's Ariane 5 flew five times in 2005. Image courtesy ESA.*

Inset: *Foton M spacecraft. Image courtesy Nicolas Pillet.*

May: *Foton*

May saw the launch of the Foton M-2 mission. This was carried out by Russia for the European Space Agency. Foton is the old Soviet space cabin used for the early Zenit series of spy satellites and for the first manned spaceship Vostok and is a spherical, recoverable cabin. Foton M-1 came to grief, exploding soon after liftoff from Plesetsk on 15 October 2002 and this was a reflight of that mission. Foton M-2 carried 385kg of experiments into biology, fluid physics, materials science, meteorites, radiation dosimetry and exobiology, all controlled from the European Space Operations Centre in Esrange, Kiruna, Sweden. ESA experts were on hand to retrieve experiments from the cabin when it came down on the flat steppe to the south of the Ural Mountains near Orenberg on 16 June. Foton M-3 is planned for 2007.

July: *the Shuttle returns to flight*

For the American National Aeronautics and Space Administration (NASA), the Return to Flight (RTF) of the shuttle was the highlight of the year. Many modifications had been made following the *Columbia* disaster of February 2003. These had taken longer than expected and the shuttle eventually returned to the skies on 26 July. The mission was commanded by Eileen Collins, with Jim Kelly as pilot. The other astronauts were Wendy Lawrence, Charles Camarda, Andy Thomas and Steve Robinson, with Soichi Noguchi from Japan. *Discovery* successfully reached the International Space Station, delivered equipment and supplies and brought several tonnes of cargo (and rubbish) back to Earth. Astronaut Steve Robinson made a spacewalk to inspect the shuttle and make minor repairs. Discovery landed in pre-dawn California on 9 August. Analysis of the mission found that debris had still fallen off the external tank during the ascent to orbit, so the shuttle was grounded again while further efforts were made to solve the problem.

October: *China's second manned spaceflight*

China became the world's third country to put astronauts into space in 2003 when Yang Liwei circled the Earth for a day. Two years later, China was ready to make another great leap forwards: this time, two astronauts (or *yuhangyuan* in Chinese) would circle the Earth for five days. The whole mission went extraordinarily smoothly. Fei Junlong, 40 and Nie Haisheng, 41, were launched aboard Shenzhou 6 on 12 October 2005. This time, the Chinese covered the take-off live from their launch site in Jiuquan in north-west China and even fitted small cameras onto the rocket so that viewers could look back to Earth and see the various stages falling away. Once in orbit, Fei Junlong and Nie Haisheng used the two cabins of the Shenzhou to carry out experiments and observe the Earth passing below. Four large tracking ships followed the mission from the Earth's oceans. They came back to Earth early on the 17 October, returning to the grasslands of northern China for a heroes' welcome. Shenzhou 7 will be the first Chinese spacewalk. Dockings and a space station will follow.

Launches by country

	2005	2004
Russia	27*	24*
USA	13	18
China	5	8
Europe	5	3
India	1	1
Japan	2	0

** Includes Sea Launch*

Above: *Discovery in Earth orbit. NASA's highlight of the year was the Shuttle's return to flight. Image courtesy NASA.*

Top: *China's piloted spaceship, the Shenzhou.*

In terms of launches, Russia remains the world's leading spacefaring nation, making more launches than the rest of the world put together. The Russian figure includes the four Zenit 3SL launches, which although it is now an international project has its roots in the old Soviet space programme. Russia dominated the annual launch figures from the 1960s, except for during the most difficult years of retrenchment after the end of the Soviet Union. Russia made a further four launch attempts that failed (see: *They failed*, below). China and Europe share third place. Historically, Europe has held the third position, but this changed in 2004 when China pulled ahead. Launches by India and Japan have always been low in numbers and generally in the order of one or two a year. There were no launches by the other spacefaring nation, Israel, in 2005.

They went up: manned launches

There were four manned space launches in 2005: two by Russia and one each by China and the United States. Fifteen people were launched into orbit: two Russians, two Chinese, one Japanese, one European (Italian) and nine Americans, including space tourist Greg Olsen. Two crewmembers were in orbit aboard the International Space Station when the year began (Leroy Chiao and Salizhan Sharipov) and they returned to Earth in April. Valeri Tokarev and William McArthur were on board the ISS as 2006 came in.

15 April	***Soyuz TMA-6***	Baikonour	Soyuz FG	**Russia**	***Sergei Krikalev*** *(Russia)* ***John Phillips*** *(USA)* ***Roberto Vittori*** *(Europe)*
26 July	***STS-114***	Cape Canaveral	Shuttle	**USA**	***Eileen Collins*** *(USA)* ***Jim Kelly*** *(USA)* ***Soichi Noguchi*** *(Japan)* ***Steven Robinson*** *(USA)* ***Andy Thomas*** *(USA)* ***Wendy Lawrence*** *(USA)* ***Charles Camarda*** *(USA)*
30 September	***Soyuz TMA-7***	Baikonour	Soyuz FG	**Russia**	***Valeri Tokarev*** *(Russia))* ***William McArthur*** *(USA)* ***Gregory Olsen*** *(tourist)*
12 October	***Shenzhou 6***	Jiuquan	CZ-2F	**China**	***Fei Junlong*** (China) ***Nie Haisheng*** (China)

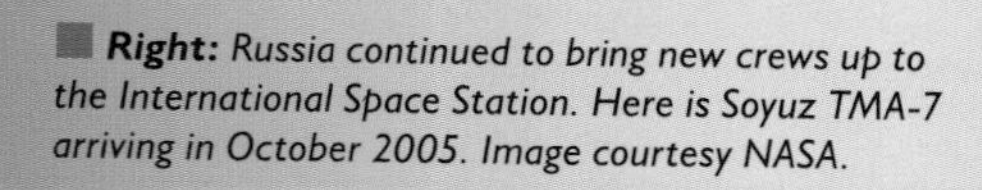

Right: *Russia continued to bring new crews up to the International Space Station. Here is Soyuz TMA-7 arriving in October 2005. Image courtesy NASA.*

The Soyuz missions changed the crew of the International Space Station, bringing a new crew up, while the old crew returned in a Soyuz already attached to the station. Roberto Vittori and Greg Olsen made week long up-and-down missions to visit the station. The shuttle visited the International Space Station but did not change the crew. Shenzhou 6 was a solo 5-day mission.

Supply missions to the International Space Station

During 2005, Russia launched four re-supply missions to the International Space Station. The Progress spacecraft is used and it is designed to carry fuel, water, equipment, experiments, food, laundry, supplies and personal items to orbiting space stations. Progress is derived from the manned Soyuz spaceship and flies entirely automatically. Progress has three series. The first, Progress, supplied the Salyut 6 and 7 space stations and concluded with Progress 42. Progress M was introduced in 1989 and was an improved version. A third variant, Progress M1 was introduced in 2000 and eleven have flown. Russia continues to operate both the M and M1 series. The M series carries greater supplies of water and because of high water demands on the International Space Station, the M version was more in use in 2005.

28 February	***Progress M-52***	Baikonour	**Soyuz U**	Russia
16 June	***Progress M-53***	Baikonour	**Soyuz U**	Russia
8 September	***Progress M-54***	Baikonour	**Soyuz U**	Russia
21 December	***Progress M-55***	Baikonour	**Soyuz U**	Russia

In addition to their normal cargo, some also carried extra items. Progress M-52 carried 50 snails to test how they would react to weightlessness. Progress M-52 also carried communications equipment to make possible a docking between the International Space Station and the European Space Agency's Automated Transfer Vehicle, the *Jules Verne*. Progress M-55 brought up Christmas and new year presents. The Progress is sometimes used to raise the height of the space station. On 10 November, for example, Progress M-54 fired for 33mins to raise the orbit of the station to 344 by 352km.

Launches by launch site

The satellites of 2005 were put into orbit from eleven launch sites. By far the busiest was the Russian launch site of Baikonour, which is located in Kazakhstan. The busiest American site was Cape Canaveral (eight), followed by Vandenberg Air Force Base, California (five).

	2005	2004
Baikonour, Russia	19	17
Cape Canaveral, USA	8	14
Kourou, French Guyana	5	3
Vandenberg, USA	5	3
Plesetsk, Russia	4	5
Jiuquan, China	4	2
Odyssey Platform, Pacific	4	2
Xi Chang, China	1	4
Sriharikota, India	1	1
Tanegashima, Japan	1	0
Kagoshima, Japan	1	0

Here is more information about the different launch sites.

Russia/Ukraine

The oldest Russian launch site is Kapustin Yar, near the Volga river, which was used for Russia's first postwar rocket tests. It was used for the launch of small satellites from 1962 to 1977 but

Above: *The Plesetsk cosmodrome. Image courtesy ESA.*

is rarely used now. Russia's best known and premier launching base is Baikonour, which, under the terms of a treaty between the two governments, is Russian territory within Kazakhstan. Its traditional name is Tyuratam. Russia's third launch base is Plesetsk in the far north, near the town of Mirny and is used principally for military launches. During the 1980s, this was the busiest spaceport in the world. Work is taking place to make Plesetsk ready for Russia's new launcher, the Angara. Russia also has a launch base at Svobodny Blagoveshchensk in the far east, used for small scientific satellites, but in 2005 the decision was taken to close the base by 2009. Submarines are also used for launches from under the Barents Sea, but here, two launches failed. In addition, the Russian/Ukrainian launch system Zenit 3SL (SL stands for Sea Launch) uses the Odyssey Platform. This is an international project, owned by companies in the United States (Boeing), Russia (Energiya) and Ukraine (Yuzhnoye). The Odyssey Platform is a converted oil rig operating in international waters in the Pacific Ocean and served by a logistics vessel whose home base is Long Beach California.

Russian/Ukrainian launches in 2005

Baikonour	**19**
Plesetsk	**4**
Odyssey Platform	**4**

United States

The United States has two main launch sites. The most famous is Cape Canaveral, Florida, the best known launch site in the world. This serves both NASA (its part is called the Kennedy Space Centre) and the United States Air Force (the Eastern Test Range). The Kennedy Space Centre will start undergoing changes in a number of years time to adapt facilities for the new launchers associated with the return to the moon. Vandenberg Air Force Base, along the hilly Pacific coast in California, is largely a military launch site and is also known as the Western Test Range. Using the Pegasus system, American satellites are put into orbit from rockets dropped from aircraft at high altitude. The aircraft used was originally the B-52 bomber, but this role has been taken over by the Lockheed L-1011, normally operating out of California. The Pegasus system was used for DART in April. A launch site has also been developed in Kwajelein Atoll for private launches.

American launches in 2005

Cape Canaveral	**8**
Vandenberg Air Force Base	**5**

Europe

Europe has one launch base, Kourou in French Guyana. It was originally a French launch base, built in 1965 when France was obliged to evacuate its first desert rocket base in Algeria, Hammaguir. The launch site was cut into the jungle and used for French launchings (Diamant rockets), the final launches of ESA's predecessor, ELDO and, from 1979, Europe's Ariane. French

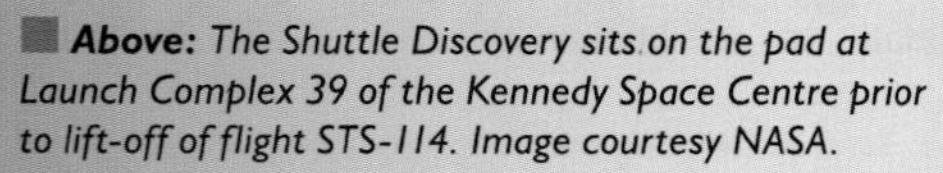

Above: *The Shuttle Discovery sits on the pad at Launch Complex 39 of the Kennedy Space Centre prior to lift-off of flight STS-114. Image courtesy NASA.*

Right: *Europe's launch site at Kourou in French Guyana. Image courtesy ESA.*

Guyana remains French territory and is part of the European Union. It is currently being expanded for the Soyuz launcher (see Laurent de Angelis: *Soyuz in the jungle,* Chapter 6).

European launches in 2005

Kourou, Guyana	**5**

China

China has three launch bases. Its original base is Jiuquan, in the high desert in north west China, used for the first Chinese satellite, Dong Fang Hong, in 1970. The site was extensively modernized in the 1990s in advance of the manned spaceflight programme, with the building of a large vehicle assembly building. The second launch site is Xi Chang in the hills of Sichuan south west China, used for sending communications satellites into 24hr orbit. Taiyuan is the third site, a military rocket base near Beijing, used for applications satellites, though it was not used in 2005. Plans have been reported for a large new launch site on Hainan island, off the south-east coast, to take the new Long March 5 rocket.

Chinese launches in 2005

Jiuquan	**4**
Xi Chang	**1**

Japan

Japan has two seaside launch sites, not far apart from one another, at the extreme rocky southern tip of the Japanese islands. Kagoshima or Uchinoura was the original site used for the small rockets that put the first Japanese scientific satellites into orbit in the early 1970s for the Institute of Space and Astronautical Sciences (ISAS). Tanegashima is a larger site, used by the Japanese National Aeronautics and Space Development Agency (NASDA) for the larger N-I, H-I and H-II rockets, including the present H-IIA. The two sites continue to follow this pattern, even though ISAS and NASDA have now been joined as JAXA.

Japanese launches in 2005

Tanegashima	**1**
Kagoshima	**1**

India

India has one satellite launching centre, Sriharikota, on the country's sandy south east coast, 100km north east of Chennai (Madras). It has two launch pads: one for the Polar Satellite Launch Vehicle (PSLV) and the other for the Geo Stationery Launch Vehicle (GSLV).

Indian launches in 2005

Sriharikota	**1**

Other launch sites: Israel uses a desert launch site called Palmachim for its Shavit launcher, but there were no missions in 2005. Brazil has a coastal launch site, Alcantara, from which it has three times attempted to launch satellites into orbit. Woomera, Australia, was Britain's main rocket launching base but following the launch of Britain's only satellite, Prospero, in 1971, it was decommissioned.

They came back: *satellites recovered*

Nine satellites returned to Earth for recovery in 2005. The space faring countries use several different recovery zones.

The main recovery zone for Russian manned spacecraft is the desert around Arkalyk, Kazakhstan, north of the Baikonour launch site. Traditionally, returning unmanned military and civilian spacecraft used to be recovered there too, but when the Kazakhs began to charge the Russians for doing so, recoveries were moved a small distance north into Russian territory south of the Urals near the town of Orenberg.

The United States used to recover its

Left: *Japan's Tanegashima launch site, with H-IIA on the pad. Image courtesy JAXA.*

manned spacecraft at sea - either in the Atlantic Ocean or the Pacific Ocean while returning military cabins were snared in mid-air by planes. The shuttle required aeroplane-type runways and three main sites were defined for shuttle landings: the desert Edwards Air Force Base in southern California; a purpose-built runway at the Kennedy Space Centre; and a military base in White Sands, New Mexico. The Kennedy Space Centre is the preferred landing site, for the landed shuttle can then be towed straight to the Orbiter Processing Facility for its next launch. In the case of STS-114, NASA hoped to bring *Discovery* back to Cape Canaveral but was waved off twice because of thunder storms and so the shuttle landed at Edwards instead. It was then placed on top of its carrier aircraft, a Boeing 747, for the return flight to Kennedy Space Centre. A number of military bases worldwide are designated as shuttle emergency runways (e.g. Spain). Although shaped like Apollo, the new Crew Exploration Vehicle is expected to return to Earth on land and for this a site is likely to be used in the high desert of the western United States. From September 2004, the Americans began to use a recovery site in the Utah desert, recovering the Genesis mission then and later the Stardust probe there in January 2006.

China has two recovery sites. Its original landing site was mountainous Sichuan, south west China, where the FSW recoverable cabins still return to Earth. For the Shenzhou manned series, a landing site is used in the flat grasslands of inner Mongolia and to the north of Beijing.

Japan has recovered one satellite, the USERS technology demonstrator (May 2003), the splashdown zone being in the Pacific near the Bonin islands. For Hayabusa, which touched down on asteroid Itokawa in November 2005, Japan will use a land site, the western Australian desert, where Hayabusa is expected in June 2010.

Recoveries in 2005

Cosmos 2410 *Kobalt*	10 January 2005	**Orenberg, southern Russia**
Soyuz TMA-5	25 April 2005	**Arkalyk, Kazakhstan**
Foton M-2	15 June 2005	**Orenberg, southern Russia**
Discovery	9 August 2005	**Edwards AFB, California**
FSW 21 (Jian Bing 4-4)	29 August 2005	**Sichuan**
FSW 22 (Jian Bing 4-5)	15 September 2005	**Sichuan**
Soyuz TMA-6	11 October 2005	**Arkalyk, Kazakhstan**
Shenzhou 6	17 October 2005	**Siziwang, Inner Mongolia**
Cosmos 2415 *Kometa*	15 October 2005	**Orenberg, southern Russia**

Comos 2410 was a Russian photo reconnaissance satellite in the *Kobalt* series introduced in 1982. Normally *Kobalt* satellites orbit for about 120 days, but this one was brought down after 107 days. Normally, returning cabins are found within hours, but this Kobalt may never have been found. Initially, it was thought that it might have been buried in deep winter snow. The fact that Cosmos 2410 made some irregular manoeuvres in orbit and was brought back early opens the possibility that the cabin did not make it past reentry.

Cosmos 2415 was a recoverable military observation satellite in the *Kometa* series introduced in 1981 and used to compile high-accuracy military topographical maps. It was the first *Kometa* for five years. Cosmos 2415 was the first *Kometa* since Cosmos 2373 of 29 September 2000 and was recovered after a standard 44 days.

The Soyuz TMA-5 cabin had been launched in October 2004 and came back with expedition 10 crew in April 2005 after its six month tour of duty on the International Space Station. Coming back with expedition 10's Salizhan Sharipov and Leroy Chiao was European astronaut Roberto Vittori after a week-long visit to the space station, called the Eneide mission.

The Soyuz TMA-6 cabin was launched in April 2005 and came back in October 2005 at the end of its six-month tour of duty with expedition 11 crew Sergei Krikalev and John Phillips. Returning with them was Space Adventures' Greg Olsen, the third space tourist. There was a pressurization problem as the Soyuz left the International Space Station, echoing the disastrous events of June 1971 when the crew of Soyuz 11 was lost. Following that catastrophe, the cosmonauts were required to wear suits from undocking to touchdown, so if the problem had worsened, they should have survived. On Soyuz TMA-6, pressure fell from 780mm to 680mm, where it stabilized. When the orbital module was cast off, the system re-sealed itself. The culprit: a foreign object, probably part of a seat buckle, caught in the hatch seal.

Discovery was the return to flight of the space shuttle, designated STS-114 (Space Transportation System 114). The shuttle should have returned to Earth on 8 August, but bad weather at Cape Canaveral forced a delay of one day. The weather at Cape Canaveral had still not improved the following morning, so the alternative landing site at Edwards Air Force Base in California was used instead. The shuttle came in to land in the dark and was followed throughout the descent by infra-red cameras.

FSW 21 and 22 were recoverable Chinese satellites in a series that dates to 1974. FSW stands for Fanhui Shi Weixing and these are cabins designed for Earth observations and microgravity experiments. FSW cabins originally orbited for only three days, but the Chinese gradually extended their operations to their current length. Chinese satellite designators are complicated, confusing and changing! The FSW series is sub-

Opposite: *After landing at Edwards AFB in California, Discovery returns to KSC in Florida on the back of its carrier aircraft. Image courtesy NASA.*

Left: *Return to flight. The crew of Discovery on STS-114. Image courtesy NASA.*

divided into FSW 0, FSW 1, FSW 2 and FSW 3 series and the programme designator Jian Bing (JB) is also used. FSW 21 and 22 are the terms most frequently applied by the Chinese themselves and are used here. Using the JB designator, the two satellites should be referred to as Jian Bing 4-4 and Jian Bing 4-5. Jian Bing 4-2 and 4-4 were used for close-look observation missions while Jian Bing 4-1, 4-3 and 4-5 were used for area-survey observations. FSW 21 (Jian Bing 4-4) returned after 27 days; FSW 22 (Jian Bing 4-5), which carried silkworm experiments, after 18 days.

They were de-orbited - *satellites taken out of orbit in 2005*

Three satellites were deliberately brought out of orbit in 2005. These were all Progress supply missions to the International Space Station. These spacecraft are not recoverable (though earlier on, some flew a recoverable capsule called Raduga). All are burned to destructive reentry over the Southern Ocean, well away from the shipping lanes so as to minimize the chance of debris impacting on boats. Normally, Progress undocks from the space station just before the arrival of a new Progress. It is then commanded to fire its engines two orbits later and crash into the Southern Ocean east of New Zealand. There were two exceptions to this pattern in 2005. First, Progress M-51 undocked on 28 February, but instead of an immediate deorbit made almost two weeks of independent flight to test orientation systems. Second, Progress M-54, which arrived at the space station in September 2005, did not deorbit in advance of the arrival of Progress M-55 in December. Instead, it was decided to keep it docked for a further three months.

Progress M-51	9 March 2005
Progress M-52	15 June 2005
Progress M-53	7 September 2005

They failed - *missions lost in 2005*

There were four space failures in 2005. Strangely, they took place on two periods of almost the same day. All were Russian, which is unusual, for the programme has a reputation for reliability, especially the small Rockot which had hitherto had a perfect record.

Molniya 3K	21 June 2005	Plesetsk	Molniya
Solar sail	21 June 2005	Barents Sea	Volna
IRDT	7 October 2005	Barents Sea	Volna
Cryosat	8 October 2005	Plesetsk	Rockot

To make things worse for Russia, two were foreign missions. The solar sail mission was a solar sail test paid for by space amateurs in the Planetary Society. The mission has sometimes been termed Cosmos-1, but this designation is avoided here for risk of confusion with the small scientific satellite Cosmos 1 launched by Russia in March 1962 and will be called Cosmos solar sail (*Solnechny Parus* in Russian). The sail was intended to be an operational test of solar sailing, an exciting, innovative mission. The actual spacecraft and sail were built by the Lavotchkin design bureau, the Russian scientific production company that built all the unmanned Soviet and interplanetary missions from 1965 onward. The rocket used was a cold war missile, the Volna, fired from an undersea submarine, the *Borisoglebsk*. The circumstances in which the Cosmos solar sail was lost remain a mystery. According to the Russians, the Volna stage stopped firing at 83sec due to a turbopump malfunction. It seems that signals were picked up from the solar sail 6 mins into the mission and some analysts believe that the failure actually took place at the final stage of entry to orbit instead.

The failure of Molniya 3K was attributed to

Above: *Progress spacecraft are normally de-orbited over the southern ocean. Here, a Progress supply ship approaches the International Space Station on 18 November 2000, bringing the Expedition One crew two tons of food, clothing, hardware and holiday gifts from their families. Image courtesy ESA.*

Opposite: *The Rockot launcher. Image courtesy ESA.*

excessive vibration in a second stage motor. This was in turn traced to a manufacturing fault in the factory, which should have been identified when it was checked before launch.

The IRDT inflatable re-entry demonstrator was launched by the submarine *Borisoglebsk*. Unlike the solar sail launch, the firing of the rocket went smoothly. It appears, though, that the trajectory was not accurate and it may have come down in the Pacific ocean, instead of landing in the Kamchatka peninsular. Two earlier inflatable tests using the Volna failed as well: 20 July 2001 (submarine *Borisoglebsk*) and 12 July 2002 (submarine *Ryazan*). Only one IRDT test has been partially successful, using the Soyuz Fregat launcher in February 2000.

The Cryosat failure was due to a software fault. The second stage should have stopped firing, dropped off and allowed the Briz KM third stage to ignite. Instead, the computer command to stop the second stage was not entered and the second stage burned to depletion. The top-heavy rocket plunged back to Earth, crashing in the Lincoln Sea, just north of Greenland, ironically the very part of the planet that Cryosat was designed to study. The Rockot launcher had been highly reliable up to this point and this was its first failure. The Russian Space Agency stressed that the hardware did not fail and the fault was human error in computer programming.

By contrast, China continues to maintain its high reputation for reliability. Shenzhou 6 was the 44th successful launching in a row.

ROCKETS OF THE WORLD

Here we look at the rockets used in 2005.

Russia

	Stages	Length	Launch weight	Payload
Proton	4	48.6m	*690,000kg*	**20,600kg**
Proton M	4	49m	*723,943kg*	**3,207kg (GTO)**
Soyuz (U version)	3	49.5m	*309,000kg*	**7,500kg**
Molniya M	4	45.2m	*305,000kg*	**1,600kg (GTO)**
Volna	2	14m	*40,000kg*	**110kg**
Rockot	3	29m	*107,000kg*	**1,850kg**
Dnepr	3	30.7m	*210,800kg*	**4,000kg**
Cosmos 3M	2	31.4m	*109,000kg*	**1,780kg**

The most used are the rockets in the R-7 series, which dates to 1953 and was the original rocket used for Sputnik in 1957. The Soyuz U series was introduced in the 1960s and the Soyuz FG version in 2000. Molniya is the four-stage version originally introduced for interplanetary probes in 1960. Cosmos 3M is a small rocket originally introduced as the Cosmos 3 in 1964. The nine-in-one launch on 27 September was the 436th Cosmos 3M.

Proton is the most powerful rocket in the Russian fleet. The rocket was introduced as a heavy-lift rocket in July 1965. Proton was used to lift Russia's space stations and space station modules into orbit and had a lifting capacity in the order of 20 tonnes. The four-stage version was used for lunar and interplanetary probes as well as for communications satellites. A more powerful version, the Proton M, was introduced in 2001 with a new upper stage, the Briz. The Anik F1R launch was the 315th Proton.

Russia's other rockets, the Volna, Rockot and Dnepr are civilian conversions of military missiles from the cold war. Volna is a submarine launched ballistic missile.

Russia/Ukraine

	Stages	Length	Launch weight	Payload
Zenit 3SL	3	48.2m	*472,600kg*	**5,896 (GTO)**

The Zenit rocket goes back to 1976, when the Soviet Union committed itself to the construction of a space shuttle to take the place of the abandoned lunar programme. The system was called Energiya and comprised four powerful side rockets. These were built in the Yuzhnoye design bureau in Dnepropetrovsk, Ukraine and were used on the only two launches of the Energiya system in 1987 and 1988. These rockets were also used in their own right as the Zenit 2 system. From 1985, the Zenit 2 was used to put Soviet military satellites into orbit. A subsequent version, the Zenit 3SL, was developed as Sea Launch (hence 'SL') as a carrier for communications satellites. This is an international project developed by Yuzhnoye (Pivdennie in Ukrainian), the Russian Energiya corporation and the American Boeing company, firing the Zenit 3SL from a converted oil platform in international waters on the equator line in the Pacific near Kiribati island.

United States

	Stages	Length	Launch weight	Payload
Delta II (7925H)	3	33.2m	*231,800kg*	**2,184kg**
Delta IV (medium)	2	47.6m	*256,300kg*	**4,211kg**
Titan IVB	3	53.2m	*915,600kg*	**21,319kg**
Atlas III	2	41.6m	*225,400kg*	**4,499kg**
Atlas V (400 version)	2	45.16m	*333,300kg*	**5,624kg**
Minotaur	4	16m	*36,400kg*	**544kg**
Pegasus	3	14m	*23,133kg*	**440kg**
Shuttle	2	47m	*2,041,000kg*	**25,401kg**

Both the Titan and Atlas series date back to military missiles in the 1950s. The Titan II was used with great success for the American manned Gemini missions over 1965-6. A more powerful version, the Titan III, was developed with side boosters for the planned Manned Orbiting Laboratory which was cancelled in 1969. The Titan 4 was the final version and one of the most powerful rockets in the world, used for a variety of civilian, military and interplanetary missions (for example, the Cassini probe to Saturn flew a Titan IVB). The Titan launched on 19 October was the last Titan 4, which will be replaced by the new Deltas and Atlases.

Atlas was the United States original intercontinental ballistic missile and a version was used to put

John Glenn into orbit in February 1962. Subsequent Atlases are so modernized and powerful that they have little in common with their predecessors apart from their name. From 1992, the Atlas was re-engineered with Russian engines, the RD-180, giving them unprecedented power. The first Russian-engined Atlas III flew in May 2000. The Atlas V was used for the Mars Reconnaissance Orbiter (see David Harland: *Arrival at the red planet*, Chapter 3).

The Delta rocket also had its origins in the cold war as an intermediate range ballistic missile, the first Delta flying in 1962. The Delta II later became the United States' most reliable medium lift launcher, being used for a wide variety of civilian and military missions to low Earth orbit. When the United States moved from launching large, heavy and costly interplanetary missions to 'faster, cheaper, better' projects, the Delta II was perfect. The Delta II comes in many versions, depending on the number of small solid rockets used on the side. The Delta IV was built in a number of versions (details of the medium version are given here).

The Shuttle's proper name in the Space Transportation System (STS) and was originally approved as a project by President Nixon. The shuttle first flew on 12 April 1981 and has since made over a hundred missions. The shuttle has probably the most unusual configuration of any rocket. The three engines of the orbiter are turned on first, burning liquid oxygen and hydrogen drawn from the huge external tank. After several seconds, while the engines build up their thrust, the two solid rocket motors on the side are ignited. The solid rockets provide enormous power for the first two minutes of the ascent and they are then dropped, falling back under parachutes into the Atlantic Ocean for recovery. The orbiter continues to fire on its three engines until it reaches orbit and the external tank is then discarded. The shuttle is the only manned spacecraft system to use solid fuel rockets and was also unique for being flown manned on its first ever flight.

The Minotaur is a recent addition to the American launch fleet, first flown in the space programme in 2000 and is an adaptation of the Minuteman missile (the Minotaur 4 is based on the subsequent Peacekeeper missile).

Europe

	Stages	Length	Launch weight	Payload
Ariane 5	2	33.8m	*748,440kg*	**6,804kg**

Europe's only operational launcher is the Ariane 5. From its formation in 1975, the main line of development for the European Space Agency was the Ariane series of rockets, derived from the L3S design. Four series were built to fly mainly commercial communications satellites into 24hr orbit, each Ariane being an improvement on its predecessor. With its use of large solid rocket boosters on the side, Ariane 5 marked a significant design departure. Although the maiden launch of Ariane 5 was a spectacular failure and although there have been subsequent difficulties, Ariane 5 has gone on to become a reliable and successful launcher capable of putting two large communications satellites into 24hr orbit. It will also be used for the European Automated Transfer Vehicle, the *Jules Verne*, to be sent to the International Space Station. Later, Ariane 5 will be joined at its launch base in French Guyana by the small European Vega rocket and the medium lift Soyuz (see Laurent de Angelis: *Soyuz in the jungle*, Chapter 6).

Above: *Atlas V sending Mars Reconnaissance Orbiter to Mars. Image courtesy NASA.*

Above left: *America's Delta II, which has sent a series of missions to Mars and the planets. This launch was of the rover Opportunity. Image courtesy NASA.*

Japan

	Stages	Length	Launch weight	Payload
M-V	3	24m	*140,200kg*	**1,800kg**
H-IIA	2	53m	*285,300kg*	**4,000kg**

Japan developed two lines of rockets. The first set was developed by the father of the Japanese space programme, Hideo Itokawa and these were very small, solid-fuelled rockets. The original series was so small it was called the 'pencil' rocket, but his rockets were responsible for the first Japanese satellite in 1970 and subsequent small scientific, lunar and interplanetary missions of the Institute of Space and Astronautical Sciences. The Mu-V is the linear descendant of these solid fuel rockets and first flew in 1997.

The H-IIA comes from the second line of Japanese rocket development. The Japanese National Astronautics and Space Development Agency required larger rockets for its applications satellites and, with American assistance, developed the N-1 rocket under licence based on the Thor system. Over time, these rockets used an increasing level of Japanese engineering, the present series being introduced with the H-II (1994) and its successor, the H-IIA which first flew in 2001.

China

	Stages	Length	Launch weight	Payload
CZ-2D	2	38.3m	*236,000kg*	**3,400kg**
CZ-2F	2	58.34m	*479,800kg*	**7,600kg**
CZ-3B	3	54.838m	*425,000kg*	**4,800kg**

China's rockets are called Long March, Chang Zheng in Chinese (another type was used briefly, the Feng Bao, or Storm). The CZ-2 series was based on the Dong Feng 5 missile and was introduced for the launch of the recoverable satellites in 1974 onward. The CZ-2 series uses nitric fuels both for the main stages and for strap-on rockets at the side. The latest and most powerful version is the CZ-2F, introduced for the Shenzhou series in 1999. China also has a Long March 4 (CZ-4), derived from the Long March CZ-2 for launches of polar satellites from Taiyuan launch centre near Beijing, but it was not used in 2005.

The CZ-3 series was introduced in 1984 for the launch of communications satellites from the Xi Chang base in Sichuan. The key departure in this series was a hydrogen-powered upper stage. There are several versions of the CZ-3: the 3, 3A and the most powerful, the 3B. The 3B suffered disaster on its first launch on 14 February 1996, later called the St Valentine's day massacre, but has since, like China's other rockets, performed reliably and there have been no Chinese launch failures since then. A new and more powerful series, of rockets, comparable to the Russian Proton, is in development. The Long March 3A, which did not fly in 2005, will be used for China's first moon probe, the Chang e (see Paolo Ulivi: *Return to the Moon: Chang e and Chandrayan lead the way,* Chapter 7).

Above: *Japan's Muses V, used to launch small interplanetary missions. Image courtesy JAXA.*

India

	Stages	Length	Launch weight	Payload
PSLV	4	38.9m	*295,000kg*	**1,360kg**

India's first rocket was the SLV, or Satellite Launch Vehicle and was a small solid-fuel rocket based on the American Scout launcher, giving way in time to a more powerful version , the Augmented Satellite Launch Vehicle, the ASLV. In the 1990s, India developed its first large rocket, the Polar Satellite Launch Vehicle, able to put applications payloads of over a tonne into polar orbit and this is the mainstay of the programme. The PSLV made its first successful flight in 1994 and will be used for India's first moon probe, the *Chandrayan* (see Paolo Ulivi: *Return to the Moon: Chang e and Chandrayan lead the way,* Chapter 7). In 2001, India introduced a much larger Geostationery Launch Vehicle (GSLV), using a Russian upper stage and this has been used to place satellites in 24hr orbit.

RECORDS

Most travelled Earth citizen

The most travelled citizen of Earth: on 16 August, Sergei Krikalev became the citizen of Earth who had spent the most time in space. He passed the record of his colleague, Sergei Avdeev, who had accumulated 747 days 14hr circling the Earth in the course of several missions and who had kept the record for a number of years. The longest duration flight on one mission remains the 438 days of Dr Valeri Poliakov on Mir, a record likely to stand until the first flight to Mars.

Sergei Krikalev's time was accumulated on several missions, dating to the Mir space station in the 1980s and he also flew on the American space shuttle. Sergei Krikalev achieved fame during the period of the collapse of the Soviet Union. Launched into orbit in May 1991, he was on board the station during the coup against President Gorbachev. In the economic chaos that followed, his return flight to Earth was delayed and he was still on the station on 31 December when the Soviet Union ceased to exist. He eventually returned to Earth in March 1992. Still with his Community Party membership card, he became called 'the last Soviet citizen'. The delay on his return - at no stage was he actually stranded - meant that he made the longest *unintended* spaceflight, five months longer than planned.

Most launched rocket

Cosmos 2415 looked like an ordinary Russian military satellite launching, a *Kometa* mapping satellite. But Cosmos 2415 also marked a milestone: the 1,700th launching of the R-7 rocket of the great designer Sergei Korolev. After the second world war, Soviet engineers brought the German A-4 rocket (better known and feared as the V-2) back for tests at the Kapustin Yar launch site on the banks of the Volga river in 1947. Called the R-1 (R for *Raket*, or 'rocket'), the Russians made a better version of their own, the R-2, followed by the R-5 (the R-3, 4 and 6 were never built). Sergei Korolev realized the limitations of the German-based technology and knew he would need something much bigger if he were

***Above:** The most travelled spacefarer, Sergei Krikalev (centre), with other members of the Soyuz TMA-6 crew, Roberto Vittori (left) and John L. Phillips (right). Image courtesy NASA.*

to get a satellite into orbit and fulfill the needs of the military for a rocket that could carry an atomic warhead. His leading design assistant Mikhail Tikhonravov took ideas from the father of cosmonautics, Konstantin Tsiolkovsky, about grouping rockets together into 'packets'. The R-7 was the first packet rocket, with a core (in Russian, blok A), with four similar stages grouped around it (blok B, V, G, D after the next letters in the Russian alphabet). Each had four engines, so a record 20 engines fired at liftoff. After several failures, the R-7 fired successfully in August 1957 on a suborbital mission and carried the first Sputnik into orbit six weeks later. The R-7 was adapted for the first manned flights and new versions were introduced for lunar and interplanetary missions (called the Molniya) and the manned Soyuz spacecraft (called the Soyuz, with sub-versions like the Soyuz U and Soyuz FG). The first of a series of new, modernized versions, the Soyuz 2, flew in November 2004 and will be the basis of the Soyuz to be launched from Kourou in French Guyana. Further modernizations are in design and there is no prospect of this obsolete, reliable and successful rocket being retired any time soon.

Most travelled rover

By 2006, five self-propelled rovers had been landed on other worlds. The first was the Russian Lunokhod, which disembarked in the Sea of Rains in November 1970. The first Mars rover was the tiny American *Sojourner,* which drove down the ramp of the Mars Pathfinder which touched down on 4 July 1997. The United States' two Mars rovers of January 2004, *Spirit* and *Opportunity* have been the most enduring of all the rovers. Designed to last only 90 Martians days or sols, they were still operating two full years after they had landed. The most travelled, though, remains Russia's Lunokhod 2 which explored the crater Le Monnier on the eastern rim of the Sea of Serenity in 1972.

Rover	Date	Landed	Duration	Distance
Lunokhod	1970	**Sea of Rains, Moon**	10 months	**10.54km**
Lunokhod 2	1972	**Crater Le Monnier, Moon**	5 months	**37km**
Sojourner	1997	**Ares Vallis, Mars**	3 months	
Spirit	2004	**Crater Gusev, Mars**	2 years+	**4km**
Opportunity	2004	**Meridiani, Mars**	2 years+	**5.5km**

Top: *The most launched rocket, the Soyuz. Image courtesy ESA.*

Above: *Still roving two years after landing: Spirit's view from the summit of the Columbia Hills. Image courtesy NASA/JPL.*

Above right: *Still the most travelled rover: Russia's Lunokhod 2.*

■ *Arianespace Flight 165. An Ariane 5 launcher blasts off into the skies of Kourou, French Guiana, above Europe's spaceport on 18 December 2004, carrying France's Helios IIA satellite.*
Image courtesy ESA/CNES/Arianespace - Service Optique CSG

2
Building the INTERNATIONAL SPACE STATION

Russia sent the first component of the International Space Station into orbit in 1998. As the tenth anniversary of the first launch draws near, the station is still far from complete. Here, *Neville Kidger* charts its progress and how much more there is still to do.

. . . a site STILL under CONSTRUCTION

and 10 years on it's far from complete

ON 2 NOVEMBER 2005, the International Space Station Programme celebrated five continuous years of human presence aboard the station. Aboard were two men – American Commander and NASA Science Officer William McArthur and Russian Flight Engineer and Soyuz Commander Valeri Tokarev. The two make up the twelvth main expedition crew to man the complex.

Since the launch of the first ISS module on 20 November 1998, aboard a Russian Proton carrier rocket, there had been one further Proton launched module and 17 American Space Shuttle flights to deliver elements and supplies to the station. In addition there had been eleven manned Soyuz ferry missions, one Soyuz assembly flight and nineteen Progress cargo missions launched to deliver crews and consumables to the permanently piloted outpost some 400km above the Earth.

With a length of some 52m and a span of 73m across two sets of solar panels mounted on top of a truss structure that gave the ISS a height of 27.5m, the complex had a mass of over 183 tonnes at the anniversary.

There have been 97 visitors onboard the station from ten countries in the first five years. Twenty-nine have lived aboard as members of the twelve station expedition crews (Table 1). Russian cosmonaut Sergei Krikalev was the only one to serve as a member of two resident crews, expeditions 1 and 11.

■ **Above:** *With the crew of the Shuttle Endeavour waiting to grapple the Russian-built Zarya control module and dock it to the US-built Unity module, an electronic still camera recorded this image of the approaching Zarya on 6 December 1998, during flight STS-88. Image courtesy NASA.*

During Space Shuttle missions and the occupancy of expedition crews, many different pairs of astronauts and cosmonauts conducted a total of 378 hours and 40 minutes of spacewalks, or Extra Vehicular Activities (EVAs) to connect modules, attached science equipment or conduct repairs. The EVAs took place from either the Space Shuttle itself or the ISS through either of the two airlocks attached to the station.

The space station, the largest international construction project in Earth orbit was, at the time of the anniversary, still several years away from final completion. When finished, the ISS will consist of science laboratories, living modules and hardware from Russia, America, Canada, Brazil, the European Space Agency members and Japan. A crew of six will conduct science investigations with the many unique facilities inside the modules. But that day is still at least five years away, as both financial problems and the Space Shuttle disaster have caused the project's completion date to be stretched out way past the original date.

Genesis of the ISS

In 1984 President Reagan laid down a new challenge for the American space agency NASA – to develop a manned space station and to do so within a decade. There followed ten years of paper studies of the Space Station "Freedom" project in various configurations, but with no actual pieces of the station being launched.

In 1993 President Clinton invited the Russians to join the effort, in part to provide work for Russian space companies to stop them working for other states which are less friendly to the United States. Together with the money and hardware from fourteen other nations the International Space Station project was born from the ashes of "Freedom". The project envisaged a first launch in 1997 and by the end of 2002 the magnificent structure would have been assembled in near-Earth orbit. The lead agency in the project was the American National Aeronautics and Space Administration (NASA). Control of the ISS is centred at the agency's Houston Control Centre with the Russian Control Centre (TsUP) in Korolev, near Moscow controlling their segment.

The statistics envisaged for the assembly of the station were staggering:

- ***A total of 37 space shuttle missions were scheduled to assemble, outfit and begin research use of the station from 1998 to 2005.***
- ***About 58 Russian launches and an ESA cargo vehicle would be needed.***
- ***About 160 spacewalks totalling 960 hours (or 1,920 man-hours), would be needed during the assembly phase for construction and maintenance.***

■ ***Above:*** *A STS-96 crew member aboard the Shuttle Discovery recorded this image of the first two ISS modules, Zarya and Unity, on 3 June 1999 with a 70mm camera during a fly-around following separation of the two spacecraft. Image courtesy NASA.*

When completed, the International Space Station would measure some 110m across the array of solar panels and the length of the combined series of science and living modules from America, Russia, Europe and Japan would amount to almost 89m. The mass would be 454 tonnes.

The crew of the station would have risen from an initial size of three people to six, working in the various modules on world-class scientific projects for the benefit of all mankind.

First launches

The FGB Control Module, called Zarya, was the first element launched. It was paid for by NASA and built in Russia. The module was intended as the central hub of the early station and was launched by Proton carrier rocket from the Baikonur Cosmodrome on 20 November 1998. The launch carried the ISS office designation "Flight 1A/R" to signify the first mission of Russia and America in the programme.

Early December saw the launch of the American Node 1, named Unity, on the Endeavour Orbiter on the STS-88 mission (Flight 2A). Unity, mounted in the payload bay of the Orbiter, was docked to the Zarya module's forward port after the Russian module had been captured by the Orbiter's Canadian-built robotic arm – Canadarm.

Three EVAs were conducted by astronauts Jerry Ross and Jim Newman to make all of the umbilical connections necessary to activate Unity. They installed handrails and foot restraint

In early 2000, the need arose to perform more work on the Zarya/Unity stack and another servicing mission - STS-101 mission (Flight 2A.2a) – was inserted into the assembly sequence and was launched on 18 May 2000. The crew of STS-101, which included the planned crew for Expedition 2, conducted more repairs and maintenance on Zarya and readied the ISS for the arrival of the Zvezda Service Module. Four new batteries, ten new smoke detectors and four new cooling fans were installed inside Zarya Control Module and a cargo boom was added outside the modules during a single EVA. Handrails were installed on Unity to aid future spacewalkers. The Shuttle Orbiter engines were used to raise the orbit of the ISS, which had no propulsive power of its own.

sockets, early communications system antennas and routing of the communications cabling from Zarya to a starboard antenna, thus beginning the planned five years of orbital assembly work to construct the ISS.

Construction of the following stages then stalled because of the ongoing funding difficulties in Russia The next module would be the Service Module, christened Zvezda. The most complex and critical element of the Russian segment, Zvezda was the planned central control post of the ISS and was to be the main crew living quarters until the arrival of the American habitation node.

At the time, the Russians were winding down operations with their Mir station but the vast bulk of the work on the Zvezda module's outfitting was still to be done. Indeed, during a visit by an American Congressman to the Khrunichev factory in Moscow only the shell of the module was actually built.

On 19 May 1999, Zvezda was shipped to the Baikonur Cosmodrome for a launch that was still at least one year away and there was the need to fly a servicing and outfitting mission to the Zarya/Unity stack. May 1999 saw the launch of Space Shuttle Discovery on the STS-96 mission (Flight 2A.1). The five person crew transferred cargo, conducted maintenance work on Zarya, including battery changeouts, installed a transmitter on the Unity node and a corrective mask on one of the targets outside Zarya, transferred and installed on the adapter section a cargo boom operator post and an adapter for its attachment that had been delivered on the Shuttle. That was to be the only visit to the station in 1999 before a firm launch date was set for the Zvezda module.

Zvezda launched

The Zvezda Service Module was launched by a Russian Proton rocket from the Baikonur Cosmodrome in the Republic of Kazakhstan on 12 July 2000. After a checkout of systems in earth orbit the Zarya/Unity stack approached and docked with Zvezda on 26 July. The combined mass of the complex amounted to 52.5 tonnes.

Now the outfitting and construction could begin in earnest. The first vehicle to be launched to the three module stack was a Russian Progress resupply ship, again from the Baikonur cosmodrome. The unmanned craft was the latest in a long series of unmanned cargo carrying craft of the series which was first launched in 1978 to the Salyut 6 space station. It carried supplies for the crew of the next Space Shuttle crew to visit the ISS.

The STS-106 (Flight 2A.2b) crew, launched on 8 September 2000 on the Space Shuttle Atlantis, activated the Zvezda module's basic systems and installed the oxygen generation system of the module (the Elektron), power supply system assemblies, computers, exercise equipment and

■ ***Above:*** *ISS as viewed from the Shuttle Atlantis on 10 September 2000 on flight STS-106. ISS was sporting a recently-arrived Progress, which appears at the top in this perspective. Next to the Progress appears the Zvezda service module, which had been delivered by a Proton rocket since the previous human visit to ISS. Image courtesy NASA.*

the toilet. An EVA was conducted to install a magnetometer on the exterior of the Zvezda module and connect cables.

STS-92 (Flight 3A), in early October 2000 carried another major part of the ISS structure. The Z1 truss segment was designed to support a large power array which would be the first major set of solar arrays on the complex. The box-like truss section also carried four Control Moment Gyroscopes (CMGs) to provide the attitude control for the ISS without the use of thruster fuel. The Z1 Truss also set the stage for the future addition of the station's major trusses which would carry more massive solar arrays and radiators. Another docking cone, called a pressurized mating adapter (PMA-3) was installed on the Unity Node, providing an additional Shuttle docking port. During an EVA, astronauts opened up a large Ku band communications antenna which provided the ISS with large volume communications and TV capability.

Manned occupation begins

The first crew to live on the ISS was launched aboard Soyuz TM-34, on 31 October 2000 from the Baikonur Cosmodrome on a Soyuz launcher. A list of the expedition crews, launch and landing dates and durations is presented in Table 1. It was intended that future crews would be carried to the station on the Space Shuttle Orbiters and that the Soyuz craft would be used as a crew escape vehicle, or lifeboat, permanently attached to the station, in which the three person crew could return to earth in the event of an evacuation of the station. The Soyuz craft would need to be swapped every six months.

The trio of experienced space travellers consisted of American William Shepherd and Russians Sergei Krikalev and Yuri Gidzenko. Krikalev became the first person to make two trips to the ISS following his flight on STS-88. The major goal of the first expedition crew was to make the station habitable for future crews by making the life support system operational and supporting three further Space Shuttle missions in what was becoming a fast-paced construction cycle.

In early December the next assembly mission took place when STS-97 (Flight 4A) delivered the first set of solar arrays and their support truss. The 15.4 tonne P6 Integrated Truss Segment was lifted and bolted to the top of the Z1 truss using the Shuttle Canadarm and astronauts working outside the Shuttle/ISS stack. The truss carried associated electronics, batteries, cooling radiator and support structure. The location was supposed to be only temporary until the rest of the huge truss segments and their solar arrays were delivered and installed in 2003.

February 2001 saw the delivery of the first major science laboratory of the growing facility – the American Destiny Laboratory. The lab module greatly increased the habitable volume of the space station. The cylindrical module was 8.5m long and 4.3m in diameter. With space for 24 racks and a roomy work area for the crew, Destiny became the centre of scientific exploration aboard the orbiting outpost. Future missions began delivering the racks.

Crew rotation

The first crew rotation using the Space Shuttle took place during March 2001 when the Expedition 2 trio of Yuri Usachev of Russia and Jim Voss and Susan Helms of the USA arrived on the STS-102 (Flight 5A.1) mission. The Shuttle also carried a pressurised cargo module – the Leonardo Multi-Purpose Logistics Module (MPLM) – one of a series of three made by the Italian Space Agency. The MPLMs are lifted from the cargo bay of the Orbiter by the Canadarm and attached to one of the docking units around the hull of the Unity node.

Delivered to the station were six system racks and the first science rack – the Human Research Facility Rack number 1 (HRF-1) designed to provide a wide range of equipment for biomedical studies of crews of the ISS. Power from the P6 solar arrays was provided into Destiny after connections were made through external plugs during the second EVA of the joint flight.

The next major addition to the ISS was the Canadian-built Canadarm2 remote manipulator which was delivered by STS-100 (Flight 6A) and activated during an EVA. Canadarm2 was the first of three pieces of the Space Station Mobile Service System, or SSMSS. Future missions will deliver the Mobile Base System -- a work platform that moves along rails covering the length of the space station -- and the Special Purpose Dexterous Manipulator, or Canada Hand.

During the activation and testing, the first major on-orbit crisis of the ISS programme occurred when all three of the Destiny Laboratory's computers failed. Control was taken over by the Shuttle Orbiter whilst the problem was fixed. The subsequent delay to the fulfilment of the STS-100 mission caused a one day delay in

the undocking of that mission. That was just one day before the arrival of the second Soyuz spacecraft carrying two cosmonauts and the first "space tourist", Dennis Tito. If the docking had been delayed further, the Soyuz might have had to wait in a separate orbit until the Orbiter had departed.

Airlock delivered

July 2001 saw the delivery and installation of the joint airlock, christened Quest. It was attached to the starboard port of the Unity node.

Two more science racks arrived with the Expedition 3 crew in August 2001 and, for the first time, an expedition crew did not see a Space Shuttle mission arrive until it delivered their replacements, Expedition 4 in December 2001. Culbertson, Dezhurov and Tyurin received a modified unmanned Progress cargo ship which left the Russian made Docking Compartment 1 (known as Pirs) attached to the Zvezda module. The ISS now had two "front doors" to space through which crews could conduct EVAs in either American or Russian suits. The trio also welcomed the third "taxi" mission to exchange Soyuz ferry ships and aided the first ESA science mission to the ISS conducted by French researcher Claudie Haigneré.

At this time, it emerged that the cost of the ISS project for NASA, already running into the multi-billion dollar bracket had risen by an unaccounted for $4 billion. The vast difference in costs between the actual and budgeted costs caused the cancellation of a planned US Habitation node and the X-38 crew return vehicle which would have been permanently attached to the station and able to return seven crewmembers in the event of an evacuation. The shuttle-like spacecraft/glider was undergoing preliminary drop test in the desert when the cancellation came. The political fallout from this event would eventually lead to the resignation of NASA Administrator Goldin.

2002: truss construction begins

The year 2002 saw more expansion of the ISS structure as a series of missions began the construction of a massive eleven-segment Integrated Truss Structure that would support the giant sets of photovoltaic solar arrays that would supply all the power needed for the complex when all the modules were finally delivered and attached.

The Expedition 4 trio helped support the initial expansion of the truss and welcomed the second space tourist, Mark Shuttleworth, in April 2002 during a Soyuz TM Taxi mission. In April 2002 the STS-110 (Flight 8A) mission delivered the central truss of the structure – called S0

■ ***Above:*** *ISS as photographed by a STS-113 crewmember on board the Shuttle Endeavour following undocking of the two spacecraft on 2 December 2002. The newly installed Port One (P1) truss now complements the Starboard One (S1) truss in centre frame. Image courtesy NASA.*

(S-Zero). STS-110 crewmembers performed four EVAs and used the shuttle and station robotic arms to install and outfit the S0. On the truss was the first "first railroad in space", the Mobile Transporter upon which the Canararm2 would ride to the farthest edges of the fully completed truss, for use.

Expedition 5 arrived on the STS-111 (Flight UF-2) mission in May 2002. The mission also delivered the Mobile Base System, or MBS. During a series of three EVAs crewmembers permanently installed the MBS onto the Mobile Transporter and replaced a wrist roll joint on Canadarm2. The three teams on the ISS during the STS-111 flight unloaded supplies and science experiments from the Leonardo Multi-Purpose Logistics Module (MPLM), making its third trip to the orbital outpost. Inside the MPLM were two more science racks for the Destiny laboratory – a NASA Express Rack and the European Space Agency's Microgravity science Glovebox.

In July 2002, the STS-112 (flight 9A) mission delivered the S1 truss segment.. The final Space Shuttle mission of 2002 was STS-113 (Flight 11A) which delivered the P1 truss segment and Expedition 6.

Columbia tragedy halts construction

In the midst of the flight of Expedition 6 a NASA science mission on the Space Shuttle Orbiter Columbia was conducted beginning on 16 January 2003. The mission was not planned to visit the ISS and was launched into a different inclination to the station (38° – the ISS orbits in a 51.6° orbit). On re-entry into the Earth's atmosphere on 1 February, the Orbiter was destroyed when searing hot plasma entered the left wing leading edge through a small hole made by a suitcase-sized piece of insulation from the external tank which had broken off during the launch.

Mission managers had decided that the impact was a maintenance issue and were not aware of the hole the foam had created in the heat shielding of the wing. The disaster was caused, in part, the Columbia Accident Investigation Board (CAIB) determined, because of schedule pressure to complete the ISS to a so-called "US core complete" status by February 2004. The remaining Space Shuttle Orbiters – Discovery, Atlantis and Endeavour - were grounded until the CAIB reported its findings and its recommendations were implemented to ensure the safe flights for the remaining ISS assembly missions.

The effect on the ISS programme of the Columbia loss was both immediate and severe. There was no way that the remaining sections of the Integrated Truss Structure could be carried into orbit with the Shuttle fleet grounded. In addition, the science modules of the European Space Agency and Japan – Columbus and Kibo –could not be flown to the ISS until the shuttle resumed flights.

A reduced crew

The crew of the ISS was reduced to just two people, a move dictated by the available consumables. The reduced crew level was still in operation at the end of 2005 and was unlikely to be changed until the full resumption of Space Shuttle missions in mid-2006 at the earliest. In this period, the Russian Space Agency became the only one of the partners able to send crews and supplies to the station using its Soyuz and Progress vehicles.

The first two-man crew, Expedition 7, consisting of Yuri Malenchenko and Edward Lu was launched from Baikonur aboard Soyuz TMA-2.

■ ***Right:*** *ISS is backdropped against a heavily cloud-covered part of Earth as the station moves away from the Shuttle Discovery during flight STS-114, following undocking of the two craft on 6 August 2005. Image courtesy NASA.*

The Expedition 6 trio returned back to Earth in Soyuz TMA-1 which had been delivered six months earlier during a taxi mission which had seen the flight of ESA astronaut Frank De Winne to conduct science operations with the new Glovebox facility. Dubbed "Caretaker Crews" the pairs of expedition crewmembers would oversee the systems of the station and conduct a reduced science programme during their six-month stays on the complex.

The Expedition 8 crew of Michael Foale and Aleksandr Kaleri conducted an EVA to place Russian commercial science payloads on the outer surface of the Zvezda module in February 2004. It was the first time that crewmembers had made an EVA with no one left inside the ISS to help with the operation of the systems of the complex and followed a lengthy debate between the American and Russian managers as to how this could be safely accomplished.

The Expedition 9 pair of Gennadi Padalka and Michael Fincke conducted four EVAs during their mission to replace a controller for the American gyroscopes which had failed and replace science payloads on the Russian modules. Two further EVAs were conducted by the Expedition 10 crew of Leroy Chiao and Salizhan Sharipov.

There was a scare on that expedition when it emerged that supplies of food on the ISS were down to very low levels forcing the men to eat less than their normal intake. The crisis (and the threat of bringing the crew home early) was averted when a Progress supply ship docked with the ISS on Christmas Day 2004 and delivered enough food for the men for several months.

Shuttle returns to flight

In July 2005, whilst the Expedition 11 pair of Sergei Krikalev and John Phillips manned the station, the Americans launched the Shuttle Orbiter Discovery on the much-awaited "Return to Flight" mission. The STS-114 (Flight LF1) was a resupply flight and docked the Raffaello MPLM for a short time to the Unity node and the Shuttle and expedition crewmembers unloaded tonnes of equipment and supplies, including a new science rack, the Human Research Facility Number 2. Several tonnes of equipment were also returned to Earth including the personal items of previous crews and Russian avionics equipment for reuse on future Soyuz and Progress ships. Three EVAs were conducted to replace a failed gyroscope on the ISS and install an equipment carrier on the Quest airlock as well as testing repair techniques on sample Space Shuttle tiles carried in the Orbiter's payload bay and removal of potentially dangerous "gap fillers" protruding from the heat resistant tiles of the Orbiter underbelly.

During Discovery's ascent into orbit, video of the external tank showed a piece of foam break away from the tank and miss the Orbiter's wing. Following Discovery's safe return to earth NASA managers again grounded the Shuttle fleet. The earliest expected resumption date for the Space Shuttle following more modifications to the huge external tanks was mid-2006.

According to the most recent manifest from NASA (see Table 2), the first mission of 2006 will be another logistics flight carrying supplies and two more science racks. STS-121 is designated Flight UF1.1 Next up will be STS-115 (Flight 12A) to deliver and install the second port truss segment and a set of solar arrays and batteries. STS-116 (Flight 12A.1) will next deliver the third port truss segment and more supplies.

In late 2005 NASA obtained permission to purchase Soyuz ferry ships from the Russians when the original agreement for use of the Russian vehicle expired in early 2006. How many more two person expedition crews will be launched to the station from Baikonur on Soyuz vehicles is uncertain.

Science summary

The ISS, NASA says, "has a unique microgravity environment that cannot be duplicated on Earth and it provides a home with 424m^3 of habitable space. It has living quarters, a galley and a weightless 'weight room,' where astronauts do aerobic and resistance exercises."

Scientific exploitation of the ISS began with a series of just four American and a larger number of biomedical, technical and Earth observations experiments during Expedition 1. Many of the Russian experiments were untended packages to measure the space environment, radiation and the effects of space on materials placed outside the station. The Russians also have a series of biomedical tests which the crews conduct on a regular cycle included blood and urine tests, measurement of the cardiovascular systems during rest and exercise and cognitive tests. The European science missions also left a number of life sciences experiments aboard the station which are used regularly.

Science research of the American Segment of the ISS grew at an ever expanding rate following

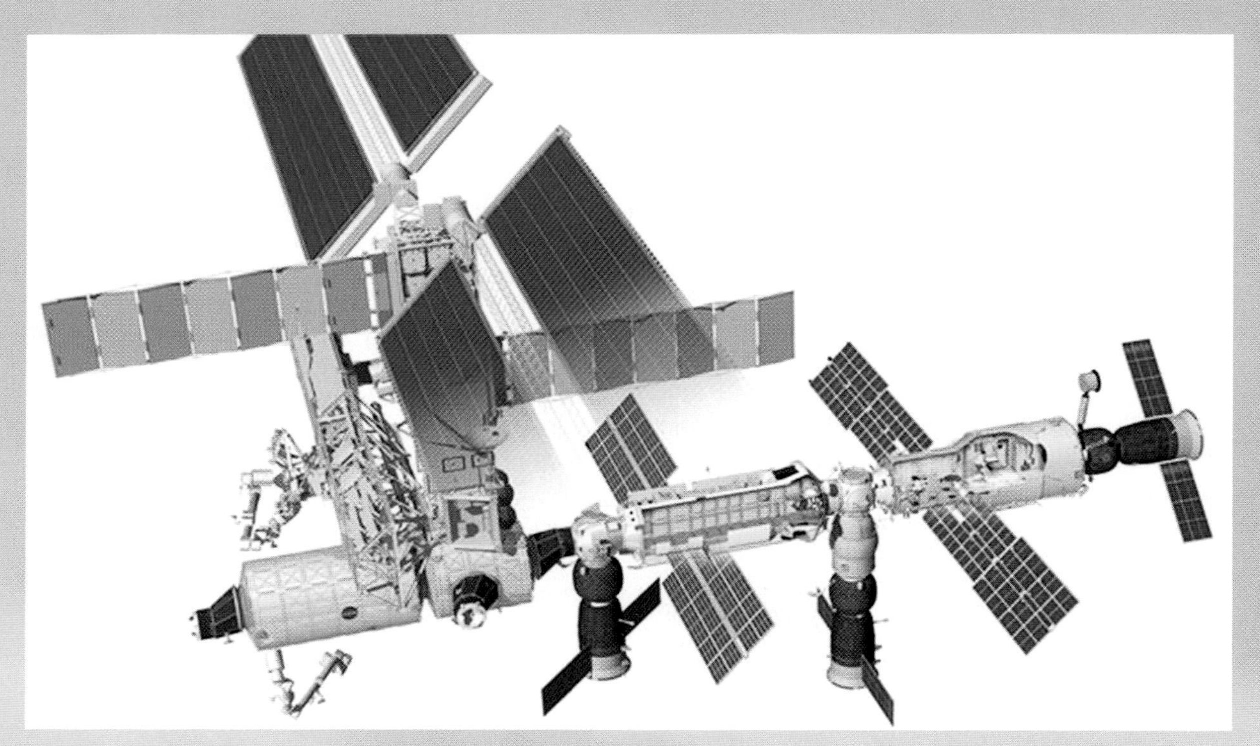

the delivery of the Destiny Laboratory module towards the end of the first expedition. Destiny was intended as the primary research laboratory for American payloads. NASA's statement said that the research was intended to support "a wide range of experiments and studies contributing to health, safety and quality of life for people all over the world. Science conducted on the station offers researchers an unparalleled opportunity to test physical processes in the absence of gravity. The results of these experiments will allow scientists to better understand our world and ourselves and prepare us for future missions, perhaps to the Moon and Mars."

The Destiny laboratory was designed to hold sets of modular racks that could be added, removed or replaced as necessary. These racks are able to contain fluid and electrical connectors, video equipment, sensors, controllers and motion dampeners to support whatever experiments are housed in them. When it arrived at the ISS, Destiny had five racks housing electrical and life-support systems. Subsequent shuttle missions delivered more racks and experiment facilities, including the Microgravity Science Glovebox, two Human Research Facility racks and five racks to hold various science experiments. Eventually, Destiny will hold up to 13 payload racks with experiments in human life science, materials research, Earth observations and commercial applications. Destiny's window – taking up one rack space - is made of optically pure glass which makes possible the taking of very high quality digital still photography and video of the Earth below.

■ **Above:** *Cutaway diagram showing the structure of the ISS when completed. Image courtesy ESA/NASA.*

The first science operations inside Destiny were performed by Expedition 2 and concentrated upon characterising the radiation and vibration characteristics of the ISS. Many small science experiments were placed in the racks and were controlled from remote locations away from the primary Payload Control Centre at NASA's Marshall Space Flight Centre in Huntsville, Alabama. Many of these research projects involved protein crystal growth and were proprietary to the organisations which had designed them. The crews of the ISS conducted regular monitoring of the experiments and performed troubleshooting, if required. An experiment growing protein crystals was left growing for 990 days.

The European Space Agency Glovebox was extensively used throughout the latter half of 2002 until a controller box broke and was returned to Earth for repair. An unmanned Progress ship flew the repaired component back to the station where it was re-installed and the Glovebox resumed work early in 2003.

After the grounding of the Shuttle fleet and the direction of NASA to pursue science which would benefit the Vision for Space Exploration alone, many of the future experiments in applied and life sciences will not happen. A planned centrifuge has been deleted from the plan. NASA plans that critical issues in human health must be resolved before humans go on missions to Mars.

The agency reports that "scientists have made great strides understanding the significant rate of bone loss by crews while in orbit and determining where that loss is occurring; vital information for long-duration missions. Because cosmic radiation is a major risk factor in human space missions, NASA scientists have used the station to test techniques to characterize the environment and generate computer models for shielding.

"Crews have trained on and experimented with medical ultrasound equipment as a research and diagnostic tool. They use a telemedicine strategy that could have widespread applications in emergency and rural care situations on Earth.

"Crews have used in-space soldering to test hardware repair techniques, providing a better understanding of fabrication and repair methods astronauts may need on long flights. Station crews have taken more than 177,000 images of Earth, providing scientists with information pertinent to scientific disciplines from climatology to geology. "

At the end of 2005 a major report stated that there would be a very limited scientific return from the ISS until the crew size was increased to six and that was not planned until 2008 at the earliest. NASA plans to operate the ISS until 2015 before the agency concentrates upon the next goal – the Vision for Space Exploration.

The new vision

In January 2004 American President George W. Bush directed NASA to retire the Space Shuttle in 2010 after completing the ISS in accordance with its international commitments and then return humans to the Moon by the year 2020 and eventually make flights to Mars. The Shuttle will be replaced with an Apollo-style capsule which will be used for crew rotation on the ISS. The vehicle is called the Crew Exploration Vehicle (CEV). NASA hopes the CEV will be "the key to making the Vision for Space Exploration a reality."

With the plan of retiring the shuttle in 2010 and the new long gap between missions it remains to be seen how many actual flights will take place before the retirement date. NASA is thought to be planning no more than about 19 missions but that figure seems in doubt and may be no more than 10. With the need to complete the assembly of the ISS under the international commitments that NASA has entered into it seems likely that one or more major pieces of the station will not reach orbit.

Apart from the completion of the ITS truss and the American node 2 there are the European and Japanese science modules still to fly. In addition, the Russian side is working towards the launch of two more modules to be used for scientific and commercial use by means of the Proton rocket in 2007. To complete the ISS by 2010 remains a hard job for NASA and the rest of the international partnership.

***Neville Kidger** is a writer on spaceflight and a regular contributor to Spaceflight magazine.*

■ **Above:** *In space, the properties of an EVA suit are of vital importance. The Russian-made Orlan spacesuits, designed by NPP Zvezda, have been highly appreciated by crews, first during missions to the Mir space station, and now during operations with ISS. Image courtesy ESA.*

Table 1: Expedition Crew Durations

Expedition	Crew	Launch	Landing	Duration
Expedition 1	*William Shepherd, Yuri Gidzenko, Sergei Krikalev*	*31 October 2000*	*21 March 2001*	*140 days 23 hrs 38 min 55 s*
Expedition 2	*Yuri Usachev, Susan Helms, James Voss*	*8 March 2001*	*22 August 2001*	*167 days 06 hrs 40 min 49 s*
Expedition 3	*Frank Culbertson, Vladimr Dezhurov, Mikhail Tyurin*	*10 August 2001*	*17 December 2001*	*128 days 20 hrs 44 min 56 s*
Expedition 4	*Yuri Onufrienko, Carl Walz, Daniel Bursch*	*5 December 2001*	*19 June 2002*	*195 days 19 hrs 38 mins 12 s*
Expedition 5	*Valeri Korzun, Peggy Whitson, Sergei Treshchev*	*5 June 2002*	*7 December 2002*	*184 days 22 hrs 14 mins 23 s*
Expedition 6	*Kenneth Bowersox, Nikolai Budarin, Donald Pettit*	*24 November 2002*	*4 May 2003*	*161 days 01 hr 14 mins 38 s*
Expedition 7	*Yuri Malenchenko, Edward Lu*	*26 April 2003*	*28 October 2003*	*184 days 22 hrs 46 mins 09 s*
Expedition 8	*Michael Foale, Aleksandr Kaleri*	*18 October 2003*	*30 April 2004*	*194 days 18 hrs 33 mins 43 s*
Expedition 9	*Gennadi Padalka, Michael Finke*	*19 April 2004*	*24 October 2004*	*187 days 21 hrs 16 mins 09 s*
Expedition 10	*Leroy Chiao, Salizhan Sharipov*	*14 October 2004*	*25 April 2005*	*192 days 19 hrs 00 min 59 s*
Expedition 11	*Sergei Krikalev, John Phillips*	*15 April 2005*	*11 October 2005*	*179 days 23 min*
Expedition 12	*William McArthur, Valeri Tokarev*	*1 October 2005*	*TBD*	*TBD*

■ ***Above right:*** Artist's impression showing the first Automated Transfer Vehicle (ATV), the European-built ISS supply ship named after French science fiction author Jules Verne, arriving at the ISS. Image courtesy ESA.

■ ***Right:*** Artist's impression showing the Japanese Experiment Module 'Kibo' arriving at the ISS. Image courtesy JAXA.

Table 2: Space Shuttle Flights and ISS Assembly Sequence

Note: Certain items are listed which are probably deleted but await a formal NASA announcement

Date	Assembly Flight	Launch Vehicle	Element(s)
No earlier than May 2006	*ULF1.1*	**Discovery STS-121**	Return to Flight test mission Utilization and Logistics Flight
Under review	*12A*	**U.S. Orbiter STS-115**	Second port truss segment (ITS P3/P4) Solar array set and batteries
Under review	*2A.1*	**U.S. Orbiter STS-116**	Third port truss segment (ITS P5) SPACEHAB single cargo module Logistics and Supplies
Under review	*13A*	**U.S. Orbiter STS-117**	Second starboard truss segment (ITS S3/S4) Solar array set and batteries
Under review	*13A.1*	**U.S. Orbiter STS-118**	SPACEHAB Single Cargo Module Third starboard truss segment (ITS S5) Logistics and Supplies
Under review	*15A*	**U.S. Orbiter STS-119**	Fourth starboard truss segment (ITS S6) Solar array set and Batteries
Under review	*10A*	**U.S. Orbiter STS-120**	U.S. Node 2

ISS U.S. Core Complete after 10A

Date	Assembly Flight	Launch Vehicle	Element(s)
Under review	*ULF2*	**U.S. Orbiter**	Multi-Purpose Logistics Module (MPLM) Utilization and Logistics Flight Crew Rotation
Under review	*ATV1*		European Automated Transer Vehicle
Under review	*1E*	**U.S. Orbiter**	European Laboratory - Columbus Module
Under review	*UF-3*	**U.S. Orbiter**	Multi-Purpose Logistics Module (MPLM) Crew Rotation
Under review	*UF-4*	**U.S. Orbiter**	Spacelab Pallet carrying "Canada Hand" (Special Purpose Dexterous Manipulator) Extended Duration Orbiter Pallet
Under review	*UF-5*	**U.S. Orbiter**	Multi-Purpose Logistics Module (MPLM) Crew Rotation
Under review	*UF-4.1*	**U.S. Orbiter**	Express Pallet S3 Attached P/L
Under review	*UF-6*	**U.S. Orbiter**	Multi-Purpose Logistics Module (MPLM) Crew Rotation
Under review	*1J/A*	**U.S. Orbiter**	Japanese Experiment Module Experiment Logistics Module (JEM ELM PS) Express Pallet
Under review	*1J*	**U.S. Orbiter**	Kibo Japanese Experiment Module (JEM) Japanese Remote Manipulator System (JEM RMS)
Under review	*ULF3*	**U.S. Orbiter**	Multi-Purpose Logistics Module (MPLM) Utilization and Logistics Flight Crew Rotation
November 2006	*3R*	**Russian Soyuz**	Universal Docking Module (UDM)
Under review	*9A.1*	**U.S. Orbiter**	Science Power Platform (SPP) solar arrays with truss POSSIBLY DELETED Multi Purpose Module (MTsM)
Under review	*UF-7*	**U.S. Orbiter**	Centrifuge Accommodation Module (CAM) POSSIBLY DELETED Crew Rotation
Under review	*2J/A*	**U.S. Orbiter**	Japanese Experiment Module Exposed Facility (JEM EF) Japanese Experiment Logistics Module - Exposed Section (ELMES) Additional Science Power Platform (SPP) solar arrays
Under review	*ULF5*	**U.S. Orbiter**	Multi-Purpose Logistics Module (MPLM) Utilization and Logistics Flight Crew Rotation
Under review	*HTV-1*	**U.S. Orbiter**	Japanese H-II Transfer Vehicle
Under review	*14A*	**U.S. Orbiter**	Cupola Express Pallet Extended Duration Orbiter Pallet

Notes:
Additional Progress, Soyuz, H-II Transfer Vehicle and Automated Transfer Vehicle flights for crew transport, logistics and resupply are not listed.

Courtesy NASA

3
ARRIVAL at the RED PLANET

Now they knew what Bunny had been busy making in the boiler room...

For the past twenty years, Mars has been the most favoured target for planetary exploration. Our knowledge of the red planet has grown enormously as a result of a series of stunning American and European missions.

Here, *David Harland* reviews what has been learned so far.

MARS exploration results

AFTER CONSOLIDATING the results of the Viking lander and orbiter missions of the mid 1970s, NASA dispatched the first of what became a wave of new planetary explorers to Mars. Mars Observer was launched on 25 September 1992 to undertake a global survey of the surface and atmosphere over the course of a local year using a high-resolution imaging system; a laser altimeter to construct a topographic map; gamma-ray and a thermal emission spectrometers to chart the composition of the surface; an infrared radiometer to monitor the annual cycle of the atmosphere; and an integrated magnetometer and electron reflectometer to seek magnetic fields. But contact with the spacecraft was lost on 22 August 1993, while it was in the process of pressurizing its propulsion system for the imminent orbit-insertion manoeuvre. This catastrophic loss was all the more significant because, for the first time, due to the high cost of the mission at a time of fiscal prudence, the agency had decided not to build a backup probe.

In response, Daniel Goldin, who had been appointed administrator in 1992, introduced what became known as a 'faster-cheaper-better' approach, by which, instead of sending a single large spacecraft with a multiplicity of instruments, a number of smaller spacecraft, each with only a few instruments, would be dispatched, two at each 'launch window'. This would not only make the programme better able to survive the inevitable occasional loss; smaller probes could be developed more quickly and, by being able to be launched on smaller rockets, more cheaply. Furthermore, since later missions would be able to carry instruments designed to follow up early findings, this sustained programme would yield better science. Prior to the loss of Mars Observer, another project had been initiated for the 1996 window, for the Discovery programme intended to demonstrate the viability of low-cost (i.e. $150m) planetary missions.

Mars Pathfinder

Mars Pathfinder was dispatched on 4 December 1996 and reached Mars on 4 July 1997. The primary engineering objective was to test an entry, descent and landing system utilizing – in turn – a heat shield, parachute, rocket thrusters and airbags, designed to enable a small lander to be delivered to a site that was rougher than a Viking could have tackled. The chosen site was Ares Vallis, an outflow channel that debouched onto Chryse Planitia, where it was hoped there would be a litter of rocks that had been swept in from a wide variety of sources. It was equipped with a stereoscopic camera and a suite of meteorology instruments.

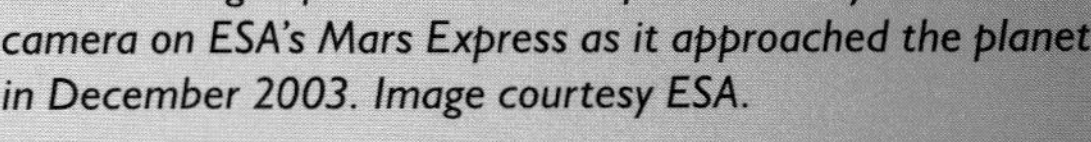
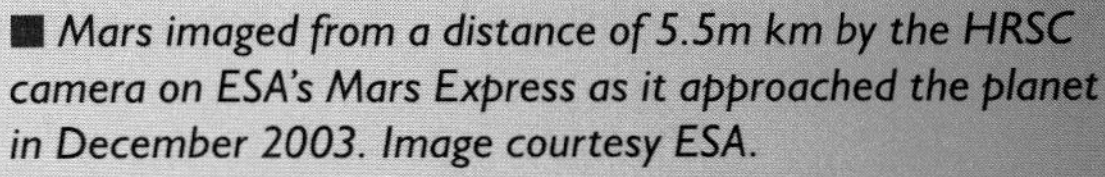

■ *Mars imaged from a distance of 5.5m km by the HRSC camera on ESA's Mars Express as it approached the planet in December 2003. Image courtesy ESA.*

Its primary payload was Sojourner, an experimental 6-wheeled rover equipped with an alpha-particle, proton and X-ray spectrometer to determine the chemical composition of the soil and individual rocks. The lander exceeded its one-month baseline mission and went on to send weather reports with better temporal resolution than had the Vikings. Hopes of monitoring the onset of the northern winter were frustrated when it fell silent on 27 September, probably because its battery froze in the night, at which time Sojourner, having far exceeded its one-week mission and analyzed eight rocks and two soil samples, was denied its communications relay. Although the results were consistent with the idea that the site marked an ancient flood, this did not prove that the planet had once had a hydrological cycle, as an intense but brief eruption of water from the surface could have occurred even in a cold and dry climate.

Mars Global Surveyor

The first spacecraft designed to pick up the research programme intended for Mars Observer was Mars Global Surveyor. Whereas Mars Observer would have used its engine to manoeuvre into its operating orbit, its successors were to use their engine to enter a highly elliptical 'capture orbit' and then penetrate the upper atmosphere at each periapsis in order to slow down and progressively reduce the apoapsis over an interval of four months, until it was economic to use the engine to circularize at an altitude of 380km. By significantly reducing the amount of propellant that the spacecraft had to carry, this 'aerobraking' technique would further reduce the mass of the spacecraft and thereby make it cheaper to launch. This initial mission was to operate for at least one local year, collecting data using upgraded versions of the high-resolution camera, laser altimeter, thermal emission spectrometer and integrated magnetometer and electron reflectometer.

A few days after launch on 7 November 1996, Mars Global Surveyor suffered damage when the deployment mechanism of one of its two solar panels became partially detached. On 12 September 1997 the craft fired its engine to enter capture orbit and when at the 54,000km apoapsis on 17 September, fired it once again to lower the periapsis into the upper atmosphere. But when the state of the damaged panel worsened early in the aerobraking campaign, the periapsis had to be raised in order to reduce the stress to an acceptable degree and because this reduced the efficiency of the braking process this phase of the mission had to be extended. In April 1998 aerobraking was suspended and the spacecraft spent five months in an orbit with a periapsis of 171km making preliminary science observations until braking was resumed on 23 September, this date having being chosen to enable the spacecraft to enter its operating orbit in optimum illumination in late February 1999.

During this period of preliminary observations, the magnetometer detected a pattern of 'magnetic stripes' in the ancient highlands, which was a serendipitous discovery as such a weak signal would have been much more difficult to detect from above the ionosphere in the operating orbit. In 2005 it was announced that the electron reflectometer had detected aurorae of electrically charged particles flowing in the magnetic field lines associated with these zones.

The primary camera (MOC) was designed to document long thin swaths of the surface at the unprecedented resolution of 1.5m per pixel and was to be used to observe specific targets. The morphological evidence of ancient lakes and river systems which it provided reinforced the hypothesis that the climate had once been conducive to the presence of liquid water on the surface. Indeed, what looks to be seepage from the steep walls of craters and channels suggests that there might still be water just beneath the surface.

Above: *The Sojourner and Spirit rovers side-by-side for comparison. Image courtesy NASA.*

The most significant 'new' data was provided by the laser altimeter (MOLA). This facilitated a global topographical map which revealed the floor of the northern basin, Vastitas Borealis, to be exceptionally flat, like the abyssal plain that forms the floor of a terrestrial ocean, and to be bounded by slopes that resemble those of Earth's continental shelves; further reinforcing the hypothesis that this basin was once an ocean. The thermal emission spectrometer (TES) was to determine the composition of the surface, in particular to confirm the presence of carbonates expected to have formed when the climate was warm and wet, but these were not evident. Olivine proved to be widely distributed and because this mineral is readily weathered in a warm and wet environment, this implied that the climate had been cold and dry for billions of years. On the other hand, the instrument found an exceptionally flat area of Meridiani Planum located in the highlands just south of the line of dichotomy to be rich in grey hematite. The reddish hue of the planet derives from its surface having been oxidized.

Hematite comes in two varieties which are chemically identical but differ in the size of their crystals. The fine-grained red hematite that eroded from rock has been distributed planetwide by dust storms. An environment involving liquid water is (usually, but not always) required to accumulate crystals of hematite into the larger grains of grey hematite. Frustratingly, therefore, the morphological evidence conflicted with the chemical evidence. As a strategy for resolving this apparent dilemma, NASA decided to 'follow the water'. Accordingly, two new spacecraft were sent to explore Mars in 1998-9: Mars Climate Orbiter and Mars Polar Lander.

Mars Climate Orbiter

In addition to an infrared radiometer (left over from Mars Observer) for profiling the atmosphere, Mars Climate Orbiter had a new imaging system capable of taking horizon-to-horizon images at medium resolution in order to monitor the weather globally on a daily and seasonal basis. The investigation was to focus on water – in particular clouds, frost on the surface and water vapour – not only to characterize the current climate but also to seek evidence that this was cyclic over a period of 100,000 years, at times being warmer than it is today. The spacecraft was launched on 11 December 1998 on a trajectory to arrive on 23 September 1999, but owing to a simple unit-conversion error by engineers, while the craft was firing its engine to enter the capture orbit, it penetrated too deeply into the atmosphere and burned up.

Mars Polar Lander

Mars Polar Lander was to use a 2m long arm fitted with a camera to excavate a trench in order to inspect the wall for fine layering that might help to determine whether the climate had undergone cyclic variations. The arm was also to supply subsurface samples to an onboard instrument that would measure the water, carbon dioxide and other 'volatiles' that were physically and chemically bound in the soil, because microbes might be able to survive in the polar regions where the presence of water would dissociate the peroxy compounds (inferred from the Viking landers at lower latitudes) that are hostile to organic material. The spacecraft was launched on 3 January 1999 and made a direct entry into the atmosphere on 3 December, but as it hovered at a height of 40m prior to landing, a design flaw prompted it to switch off its engine prematurely, so it crashed.

Mars Odyssey

As the loss of Mars Climate Orbiter had been a procedural error, it was possible to dispatch the next planned orbital mission, but the proposed lander was deleted while surface activities were reassessed. As the next window occurred in 2001, the new orbiter was named Mars Odyssey in homage to Arthur C. Clarke's novel *2001: A Space Odyssey*. It was launched on 7 April 2001, arrived on 24 October, completed its aerobraking in January 2002 and started its primary mission, which was to last at least one local year. It pursued the 'follow the water' theme using two instruments to study the surface. The gamma-ray spectrometer (the final instrument from Mars Observer) was to measure the abundances of 20 elements in the topsoil at a spatial resolution of 300km – sufficient to give a sense of the general character of the planet.

The instrument had been augmented with neutron detectors with which to sense hydrogen, indicating the presence of either hydrated minerals or water-ice in the uppermost metre of the ground. The spacecraft also had a new instrument: a thermal emission spectrometer designed to detect carbonates, silicates, hydroxides, sulphates, oxides and hydrothermal silica in the topsoil at abundances of 10%. This was integrated with a medium-resolution optical imager, to correlate the spectral data with the associated landforms. This instrument was to make a global mineralogical map with a resolution of 100m per pixel, to identify structures associated with the action of water, in order to provide further insight into the past climate.

Mars Exploration Rovers

In April 1999, NASA began field trials of the rover for the Mars Sample Return mission. The rover's task would be to analyze rocks *in situ*, drill out cores and put the most interesting samples in the return capsule. After the losses of Mars Climate Orbiter and Mars Polar Lander, the Mars Sample Return mission was placed 'on hold' and it was decided to use the airbag system to deliver a rover to conduct an independent mission to pursue the 'follow the water' strategy by seeking 'ground truth' to supplement the orbital remote sensing. For redundancy, it was decided to send two rovers. As robotic field geologists, they were to have tools to investigate the physical form, chemistry and mineralogy of individual rocks in order to reveal whether they were formed in, or were altered by, liquid water.

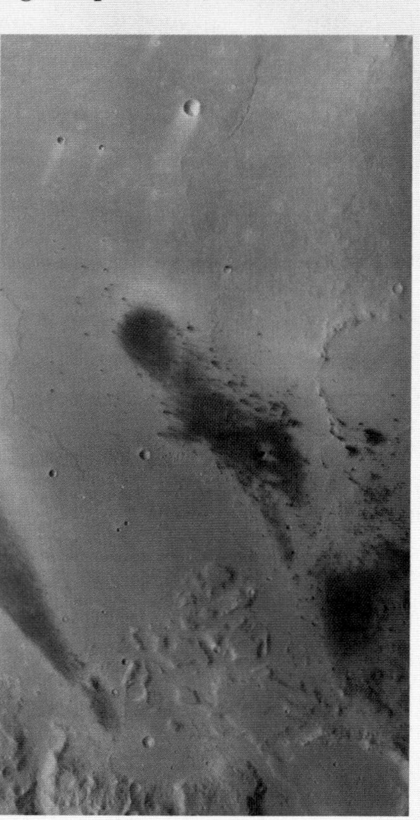

The scientific suite, known as Athena, was provided by an international team led by Steven Squyres of Cornell University. To preclude the perspective that had resulted from Sojourner's camera being so close to the ground, the cameras on the mast of the new vehicles provided a viewpoint equivalent to that of an upright human being. The panoramic stereoscopic camera had a resolution matching 20/20 eyesight and was equipped with filters for visible and near-infrared wavelengths. The monochrome navigation camera was of lower resolution, but could record a wide arc using fewer frames and thus put less demand on the limited downlink capacity. The third instrument to exploit the mast was a miniaturized version of the thermal emission spectrometer on Mars Global Surveyor. Its sensor produced a mosaic of 'false colour' circular spots that were later superimposed on an image to provide context. The longer this 'stared', the greater was its signal-to-noise ratio. Each spot provided a spectrum in 167 wavelengths. It was to assist in the selection of rocks for individual sampling using the tools on a short robotic arm capable of motions involving five degrees of freedom.

Above: *The centre of the 160km diameter Gusev crater from orbit imaged by the HRSC camera on ESA's Mars Express. Image courtesy ESA/DLR/FU Berlin (G. Neukum).*

The arm carried a microscopic imager with a resolution matching a hand lens, to assist in identifying rocks formed in water, features of volcanic and impact origin and veins of minerals left by the presence of water. For soils, it would show the sizes and shapes of the grains and yield insight into erosional processes. Like Sojourner, the new vehicle had an alpha-particle X-ray spectrometer to detect the principal rock-forming elements. Because iron interacts strongly with liquid water, a Mössbauer spectrometer was to investigate iron-bearing minerals. As the rocks examined by the spectrometer on Sojourner had proved to be coated with a rind of weathered material, the new vehicle had a rock abrasion tool (RAT) with which to brush dust off a rock and grind into its surface to expose a circle 5cm in diameter and several millimetres deep.

The engineering requirements of the landing sites were that they be located in the equatorial zone for maximum illumination of the solar panels; at low elevation for the greatest parachute braking; and not so rocky as to inhibit driving. One site was selected for its morphological character and the other on the basis of chemical remote-sensing. The first rover would be sent to Gusev, a 150km diameter crater whose southern rim had been breached by Ma'adim Vallis, a 900km long channel running off the southern highlands, seemingly forming a lake that left a deposit of sediment on the floor of the crater. The second rover was to set down on Meridiani Planum to determine whether the grey hematite that had been spotted by Mars Global Surveyor

was present in layers of sediment laid down in a lake, in veins resulting from the alteration of pre-existing rocks by a hydrothermal system, or, indeed, the result of another process not involving water.

MER-A, named Spirit, was launched on 10 June 2003 and arrived on 3 January 2004, landing about 300m southwest of a 200m diameter crater that was later named Bonneville. It relayed its transmissions via Mars Global Surveyor and Mars Odyssey. The first imagery revealed a generally flat rock strewn plain which prompted Squyres to say, "If you'd asked me ahead of time what's a dry lake bed on Mars going to look like, I'd have said a lot like this!" On sol 12 (a 'sol' is a Martian day) Spirit drove off its lander, halted and spent a few days analyzing the soil. On sol 15 it moved into position to inspect its first rock, named Adirondack, but on sol 18 was disabled by a software issue that kept it out of action for over a week and it was sol 33 before it resumed its examination of the rock, which proved to be basaltic.

Since the plain appeared to be volcanic, Spirit was directed to drive to Bonneville in the hope that the lava flow was sufficiently thin for the crater to have penetrated to the putative lacustrine material, exposing the stratigraphy in its inside wall and depositing the most deeply excavated rocks on its rim. Setting off on sol 37, the rover examined a number of rocks *en route* in order to conduct a radial survey through the blanket of ejecta that surrounded the crater. When it reached the rim on sol 66 and returned a panoramic view of the crater's floor, Bonneville was seen to be degraded with no exposures to suggest that it had excavated anything other than volcanic rock. Spirit drove anticlockwise around the rim, studying rocks, some of which proved to have undergone a degree of chemical alteration in the presence of water – although only in quantities that could be explained in terms of coatings of frost.

On sol 87 Spirit left Bonneville and set off towards a group of hills, named the Columbia Hills, just over 2km to the southeast, the tallest of which, Husband Hill, rose 100m above the plain. It was hoped the hills were *stratigraphically lower* than the lava and therefore a remnant of the putative lake bed – a not implausible theory, considering that the Gusev's wall stands several kilometres above its floor and the lake might at times have been of considerable depth.

As the rover was at the end of its nominal 90-sol life there was little expectation that it would reach its objective. Nevertheless, to the team's delight, Spirit did reach the base of the West Spur of Husband Hill on sol 156 and the contact between the two geological units proved to be very sharp, with the composition of the soil changing significantly in the span of only a few metres. Slowly ascending, Spirit finally reached the summit of Husband Hill on sol 581 (21 August 2005). On the way, it found only rocks of volcanogenic origin – quite possibly an airborne ash fall or a pyroclastic flow from Apollinaris Patera, a large volcano 250km to the north – which, although significantly chemically altered in the presence

Above top: *View of the martian surface obtained by the Mars Exploration Rover, Spirit, shortly after its landing within Gusev crater. Image courtesy NASA/JPL.*

Above: *Panorama of Gusev crater obtained by the Mars Exploration Rover, Spirit, from a vantage point on the Columbia Hills. Image courtesy NASAJPL.*

of water, had not been laid down in water. After exploiting its elevated location to produce an awe-inspiring panorama, Spirit set off down the eastern flank to explore the 'inner basin' of the Columbia Hills.

As regards the issue of the absence of carbonate, this had been resolved on the opposite side of the planet by its twin, the MER-B rover, Opportunity, which had landed on 24 January 2004, while Spirit was disabled by its computer glitch. By a remarkable piece of luck it had rolled into a small crater, named Eagle, which had finely layered bedrock in its wall. After departing its base, the rover conducted its preliminary analysis of the soil, which was very different to that seen at other sites, being made of exceptionally fine-grained dark red material mixed with small grey spherules which, it was later proved, contained the hematite.

Opportunity drove to the right-hand end of the outcrop on sol 12, where it inspected a rock named Stone Mountain, which proved to be a stack of layers, each only a few millimetres thick. "Embedded in it, like blueberries in a muffin, are little grey spherules," reported Squyres, thereby coining the term by which the spherules would become popularly known. They were seen in various stages of eroding from the fine-grained matrix, presumably as a result of wind action. When the composition of the outcrop was analyzed by the spectrometers the rock proved to be rich in sulphur.

Above top: *View of "El Dorado" dune from the Mars Exploration Rover, Spirit. Left of centre, the range of distant hills on the horizon marks the southern rim of the Gusev crater. Image courtesy NASA.*

Above: *Heading south on the "Erebus Highway", taken on 16 September 2005 by the Mars Exploration Rover, Opportunity. Image courtesy NASA.*

The rover then spent the following week working along the exposure, documenting it at maximum spatial and spectral resolution. On sol 26 it returned to the centre, where it spent 10 days making a detailed study of two exposures in a section named El Capitan. At a press conference on 3 March it was announced that the sulphates *must* have been formed by a process involving liquid water and the site had been "drenched" with water, but it was not yet possible to say whether this water stood on the surface or percolated underground.

There were multiple lines of evidence. Firstly, there was the sulphur in the form of salts, which made up 50% of the rock. There were also elements that can produce chloride and bromide salts. But the clincher was the presence of jarosite, a potassium–iron sulphate whose crystalline structure includes hydroxyl, whose formation *requires* the presence of water since it is created by the chemical alteration of basalt by acidic aqueous sulphate.

The fact that the aqueous fluid was *acidic* was a major discovery. As regards the blueberries, the microscope had shown them to be distributed throughout the rock, as opposed to being limited to distinct layers – as would have been the case if they had originated elsewhere, been deposited on the surface and then buried by later sedimentation. The fact that linearities in the surrounding matrix continued *through* the blueberries meant they had formed *in situ*. As water rich in iron-bearing minerals percolated through the rock, the minerals had precipitated and accreted around an irregularity, such as a grain of sand within the sulphate. The final line of evidence was that the rock was riddled by randomly oriented disk-shaped 'voids' which, on Earth, indicate where crystals that formed in rock which was exposed to briny water were subsequently either weathered out or dissolved by less salty water. In this case, the geometry of the voids was suggestive of evaporite minerals.

Meanwhile, the rover had studied several other parts of the outcrop. "We think we're parked on what was once the shoreline of a salty sea," announced Squyres on 23 March, in launching the press conference to report the analysis of the very fine layering in the rock Last Chance. The study was based on a stereoscopic analysis of mosaics

mosaics of images taken by the microscope. In addition to crossbedding where the layers varied in thickness along their length, indicating erosion during deposition, there were tell-tale curves called festoons that form when flowing water sifts loose sediments on the surface. In this case, the shapes indicated that the water had been at least 5cm deep and had flowed at the gentle rate of 10 to 50cm per second. A second line of evidence was the presence of chlorine and bromine, which indicated the rock was once immersed in a salty fluid that evaporated. This implied a shallow sea that sometimes dried up, leaving behind a salt flat, or playa. The earlier finding that the rocks had been drenched in mineral-rich water had not resolved whether the water was present as the rocks formed, or had altered them after their formation, possibly underground; but this new evidence indicated that they had formed *on the surface*.

"NASA launched the MER mission specifically to check whether at least one part of Mars ever had a persistently wet environment that could possibly have been hospitable to life," pointed out James Garvin, who was the agency's chief scientist for Martian exploration. "Today we have strong evidence for an exciting answer – yes!"

The fact that the hematite covered an area the size of Oklahoma indicated that a *large amount* of water had been involved. It was apparent from the light-toned rims that characterized the craters on the hematite plain that the outcrop in Eagle crater was regional. An inspection of a larger crater should provide insight into the deeper structure of the plain. Fortunately, there was such a crater 750m to the east, named Endurance and on sol 70 Opportunity set off for it. On cresting the raised rim on sol 95, it made a panorama of the interior that showed much of the southern wall to be a 10m tall near-vertical exposure, dubbed the Burns Cliff.

After the rover had sampled rocks on the rim directly above this cliff, it entered the crater on sol 133 at a point on the southwestern wall where the slope averaged 25° and, working its way slowly down, it drilled a series of holes to sample the exposure – just as a terrestrial field geologist would do. The uppermost part of the wall matched that seen in Eagle crater. Below it were several layers in which the sulphates lacked voids and water ripples, suggesting that after it had been laid down in water the material was homogenized, possibly by the wind stirring it up during dry epochs.

Further down was another preserved layer. Beneath that was a compacted sand dune. The layering in the wall of Endurance increased the extent of the vertical section by a factor of 10 over that available at Eagle and proved that although a considerable depth of rock had been altered by the presence of water, there had been substantial dry epochs. As Squyres put it, Meridiani had been "not a deep-water environment but more of a salt flat, alternatively wet and dry".

On sol 184, by which time Opportunity was about 22m into the crater, it turned right to start a drive that would enable it to take a panoramic view looking *up* at the layering in the Burns Cliff, but progress was slow due to slippage on the slope and on sol 271 it was decided to call a halt. After taking the panorama, on sol 295 the rover turned around and, to the team's delight,

Above: *ESA's Mars Express spacecraft approaching Mars. Image courtesy ESA.*

on sol 315 managed to scramble out of the crater. Although, like its sibling, Opportunity had far exceeded its nominal 90-sol mission, it was still in excellent health and so was directed to drive south off the hematite plain onto the 'etched terrain' beyond, in the hope that it would be able to reach the 1,500m diameter crater Victoria, 5km distant, in search of an even deeper exposure.

The surprising discovery that the aqueous fluid in which the sulphate was laid down was acidic offered a resolution of the 'carbonate paradox'. If an early ocean was acidified by iron and sulphur, then it would have had a very different chemical evolution to that of our own oceans. In particular, because carbonates would have been unable to precipitate from an acidic solution, it would not have left sediments such as occur on Earth.

Mars Express

Having supplied several instruments for the ill-fated Mars '96, the European Space Agency decided in 1999 to build its own spacecraft in an effort to recover some of this science. This "fast, flexible and cheap" mission would share technology with the Rosetta cometary probe and be equipped with a high-resolution colour imager, an infrared spectrometer for mineralogical mapping, two spectrometers to measure the atmospheric composition on a local scale in order to study circulation patterns and an instrument to investigate how the solar wind interacts with the atmosphere. In addition, it had a long-wavelength radar able to penetrate the ground to a depth of 5km in search of subsurface structure (in the same way as geologists prospect for water, oil, rock layers, or subsurface faults) and a lightweight landing probe named Beagle 2.

Mars Express was launched on 2 June 2003 and, as it approached the planet, on 19 December it released Beagle 2 to enter the atmosphere on 25 December, but the signal to report a successful landing was not received. This was a pity, because the lander was to inspect rocks for minerals indicative of the presence of liquid water in the past, seek carbonaceous structures left by organisms that were living in that water, measure the ratio of the carbon isotopes (a 'biomarker' test that is used to investigate the earliest terrestrial life) and seek out-of-equilibrium gases indicative of extant life in the shallow subsurface. Later that day, the main spacecraft fired its engine to enter an elliptical capture orbit, which it circularized using aerobraking and achieved its operating orbit in late January 2004. The loss of Beagle 2 became all the more frustrating when one of the orbiter's spectrometers confirmed studies by terrestrial observers which indicated there to be methane in the atmosphere that might be of biogenic origin.

In addition to confirming the crustal magnetic fields first noted by Mars Global Surveyor, Mars Express detected hydrated minerals (particularly sulphates) in dark dunes around the north pole, in Valles Marineris and a large section of Meridiani Planum – the latter enabling the results of the Opportunity rover to be extrapolated into a broader context. Most intriguing was the discovery of what has been dubbed the Athabascan 'ice sea'. A vast outflow of water from the Cerberus Fossae on the southeastern flank of the Elysium rise formed Athabasca Vallis as it drained south. Downstream is an 800km wide area patterned with irregular plates bearing a striking resemblance to a terrestrial ice floe, which suggests that as the water ran downslope it started to freeze and the floating pack-ice broke up into irregular rafts. Being located near the equator, the ice ought to have melted and the water sublimed, but the fact that it did not indicates that it was soon covered, most likely by ash originating from the volcanic Elysium province (the geothermal heat of which had prompted the

■ ***Above:** The command to deploy the second MARSIS boom was given to ESA's Mars Express spacecraft on 14 June 2005. Image courtesy ESA.*

flood) in a deposit that was thick enough to insulate the ice and yet not so thick as to mask the surface relief. It has been estimated that this ice, which covers an area of 750,000km^2 to a depth of 45m, formed within the last 5m years.

The deployment of the MARSIS long-wavelength radio-frequency antenna was placed 'on hold' when tests of a yet-to-be-launched spacecraft fitted with a similar deployment system raised the prospect of the segmented booms 'thrashing' as they unfolded, striking and damaging the vehicle. When the first 20m dipole boom was deployed on 4 May 2005, one of its 13 segments failed to lock into place, but when the craft was rolled to allow the Sun to heat the boom the mechanism locked. The second dipole boom was deployed without incident on 16 June, as was the 7m monopole boom the following day. In daylight, the instrument 'sounds' the ionosphere and at night it probes the subsurface. The early results established that deep beneath the water-ice that lies on the surface at the north pole there is a great deal of ice, but as yet there are no indications of aquifers of liquid water.

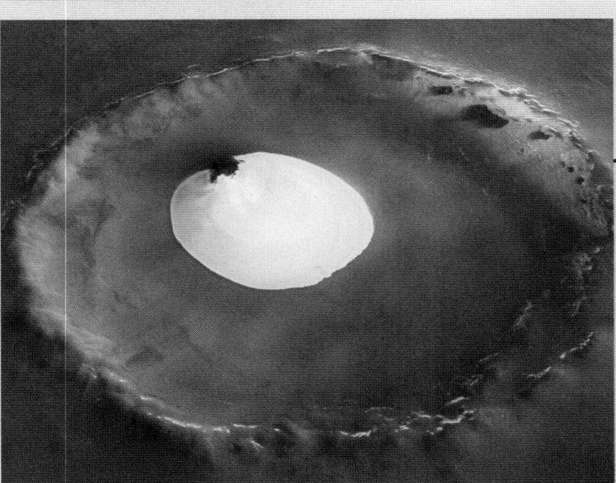

Mars Reconnaissance Orbiter

NASA's Mars Reconnaissance Orbiter was launched on 12 August 2005 and mid-course corrections on 27 August and 18 November refined its aim to arrive at Mars on 10 March 2006, when it will use its rocket engine to enter a near-polar elliptical orbit that will be circularized by aerobraking over the following six months. In contrast to the recent series of 'faster-cheaper-better' orbiters sent to Mars by NASA, each of which had just a few instruments, Mars Reconnaissance Orbiter was to undertake a multi-faceted programme to follow up the findings of its predecessors. Specifically, it was to:

- ***image selected sites at 1m resolution, particularly where stratigraphy is in outcrop or terrain has been shaped by water erosion, using the most powerful camera ever placed into orbit around the planet;***
- ***investigate selected sites in search of minerals diagnostic of water, using a spectrometer with a higher spatial resolution than previous instruments;***
- ***profile the atmosphere, using an improved form of the infrared radiometer flown (and lost) on both Mars Observer and Mars Climate Orbiter;***
- ***monitor the weather on a daily and seasonal basis, in order to further define the present climate, using an improved form of the colour-imaging system on Mars Climate Orbiter;***
- ***probe to a depth of several hundred metres in search of subsurface structures, particularly in the polar regions using a sounding radar;***
- ***use a 3m diameter high-gain antenna to provide a data-rate 10 times that of any previous orbiter;***
- ***examine potential landing sites for future missions; and***
- ***serve as an orbital relay.***

Above left: *This close-up view of Mars' north polar ice cap from the HRSC camera on ESA's Mars Express shows layers of water ice and dust for the first time in perspective. Here we see cliffs which are almost 2 km high, and the dark material in the caldera-like structures and dune fields could be volcanic ash. Image courtesy ESA/DLR/FU Berlin (G. Neukum).*

Below left: *Perspective view of an unnamed crater on Vastitas Borealis, a broad plain that covers much of Mars's far northern latitudes. The circular patch of bright material located at the centre of the crater is residual water ice, imaged by the HRSC camera on ESA's Mars Express. Image courtesy ESA/DLR/FU Berlin (G. Neukum)*

Phoenix Mars Scout

After the loss of Mars Polar Lander in 1999, a sister craft scheduled for 2001 was initially cancelled and then reassigned as the first mission in the Mars Scout series. Continuing the 'follow the water' strategy, the Phoenix lander will be launched in August 2007 and land in May 2008 in the northern hemisphere, just outside of the seasonal cap where the neutron spectrometer on Mars Odyssey showed there to be near-surface ice. It will have improved versions of some of the instruments lost on Mars Polar Lander and others from the 2001 mission. A 2m long arm will dig a trench to supply sub-surface samples to an instrument to measure the 'volatiles', in particular water and carbon dioxide physically and chemically bound in the soil, after which meteorology instruments will monitor the local environment.

Current situation

In 2006, Mars Global Surveyor, Mars Odyssey and Mars Express were all still undertaking remote-sensing in orbit, Spirit was descending the eastern flank of Husband Hill and Opportunity was progressing across the 'etched' terrain towards the large crater Victoria. Mars Reconnaissance Orbiter was *en route* to the planet, Phoenix Mars Scout was being prepared and NASA's long-term goal was to send a human mission to Mars.

Above right: *The MARSIS instrument on ESA's Mars Express spacecraft will enable scientists to take a closer look at the extent of water and water-ice on Mars and how deeply it is buried beneath the surface of Mars. Image courtesy ESA.*

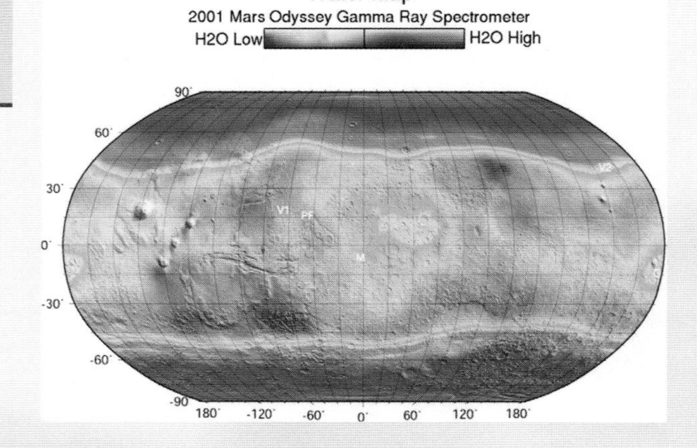

Below right: Scientists used the Gamma Ray Spectrometer aboard NASA's Mars Odyssey spacecraft to produce a global water map of Mars. The areas with the highest concentrations are at or near the martian poles. Image courtesy NASA/JPL/University of Arizona.

Dr David M Harland *gained his BSc in astronomy in 1977 and a doctorate in computational science. He has subsequently taught computer science, worked in industry and managed academic research. In 1995 he retired and has since published many well known books on space themes.*

Further reading

David M Harland:
Water and the Search for Life on Mars. Springer–Praxis, 2005.

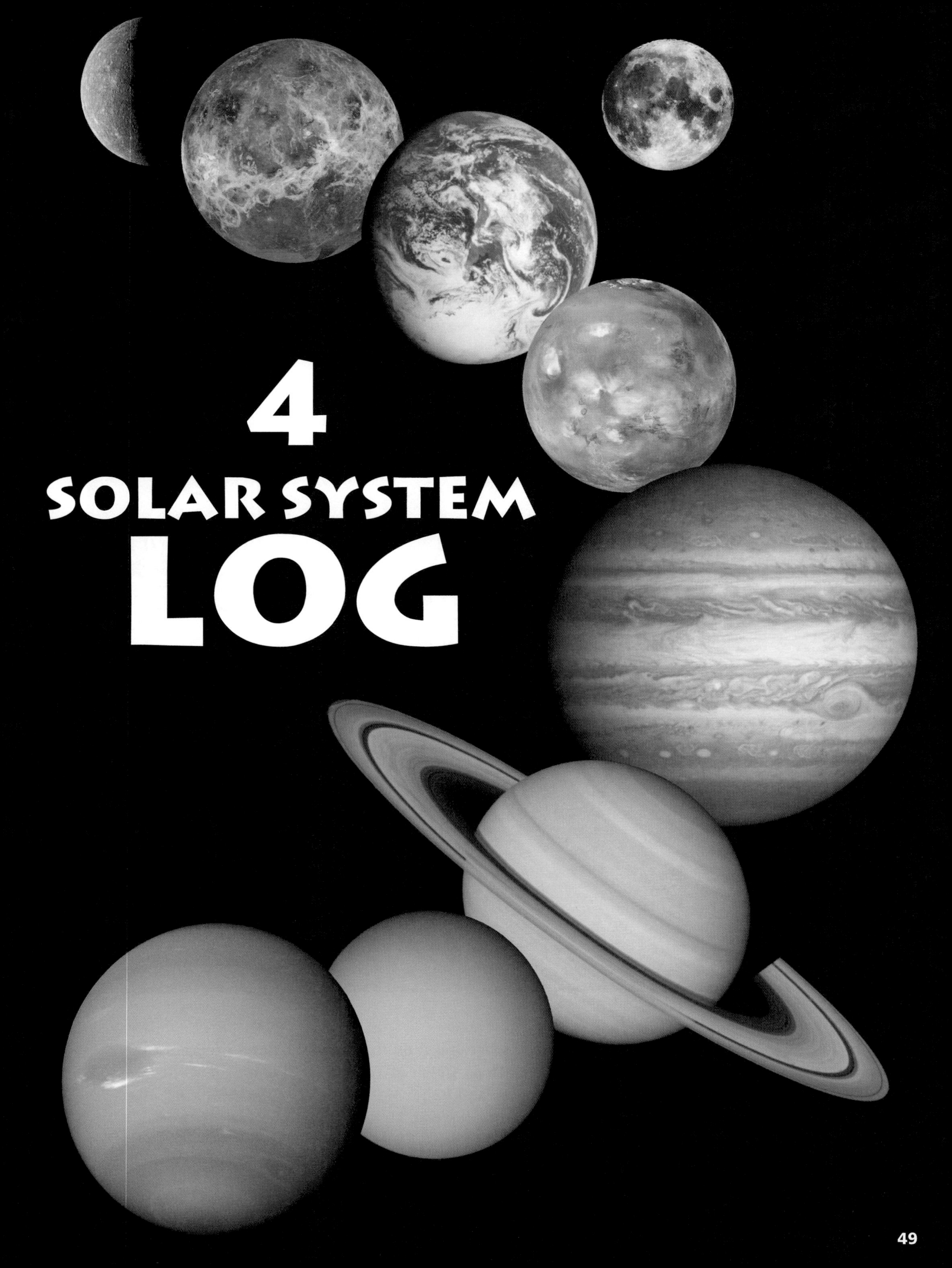

4 SOLAR SYSTEM LOG

4.1 Looking at MERCURY...

Bunny had been told Mercury was a small planet: but it looked pretty big right now!!

For the first time in thirty years, a spaceship is on its way to the Sun's innermost planet. Here, *Sean Solomon* of the Department of Terrestrial Magnetism at the Carnegie Institution of Washington and *Ralph McNutt* of The Johns Hopkins University Applied Physics Laboratory describe Mercury's complex features and their hopes for what the MESSENGER mission will bring.

The MESSENGER Mission to MERCURY

MERCURY, THE innermost planet of our Solar System, has long been a neglected target of spacecraft exploration. That situation changed in August 2004 with the launch of MESSENGER, shorthand for the National Aeronautics and Space Administration's MErcury Surface, Space ENvironment, GEochemistry, and Ranging mission (Fig. 1). After a long and circuitous route through the inner Solar System MESSENGER will become the first probe to orbit Mercury in March 2011.

The history of exploration of Mercury prior to MESSENGER includes only a single spacecraft. Mariner 10 flew by the planet three times in 1974 and 1975. Each flyby was separated by two Mercury "years" – two revolutions of Mercury about the Sun. Mercury is in a rotational state unique in the Solar System, in that the planet's spin period is precisely two thirds of the rotational period. As a consequence the solar day on Mercury — the time between successive passages of the Sun overhead — is equal to two Mercury years. Mariner 10 therefore saw the same side of Mercury lit by the Sun during each of its three close encounters, and more than half of Mercury's surface was never imaged. The images of the surface that Mariner 10 did obtain stimulated arguments about the planet's geological history that continue to the present, and other discoveries by Mariner 10 raised many questions still not answered.

Above, Figure 1: *The MESSENGER spacecraft left Earth at 2:15:56 am EDT on 3 August 2004 aboard a Boeing Delta II rocket launched from Cape Canaveral Air Force Station, Florida. (Image courtesy NASA and The Johns Hopkins University Applied Physics Laboratory.)*

Mercury's anomalous density

Even before the Mariner 10 mission, it was known that Mercury is unusually dense. After correcting for the effects of self-compression by interior pressure, the "uncompressed" density of the material inside Mercury is substantially higher than that of any of the other planets. Because Mercury, like the other inner planets, is composed of rock and metal, the high density implies that the mass fraction of metal occupying a central core in Mercury is at least 60%, a fraction twice as high as that for the Earth (Fig. 2).

■ ***Above, Figure 2:** On the basis of its bulk density, Mercury must have a central core consisting mostly of iron metal and occupying a fraction of the planetary interior much larger than that for Earth's core. Earth has a solid inner core and a fluid outer core, shown to approximate scale; Earth's magnetic field is sustained by a hydromagnetic dynamo in the outer core. The nature of Mercury's core and the origin of the planet's magnetic field remain to be determined. (Image courtesy NASA and The Johns Hopkins University Applied Physics Laboratory.)*

Mercury's high metal fraction must date from early in Solar System history when the inner planets were assembled from material within the disk, or nebula, of dust and gas that surrounded the young Sun. One hypothesis is that the material of the innermost nebula, from which Mercury was later predominantly accreted, was enriched in metal because the lighter silicate grains were preferentially slowed by interaction with the nebular gas and tended to fall into the Sun. Another hypothesis is that after Mercury accreted to full planetary size and a central metal core differentiated from a silicate shell, the silicate fraction was partially vaporized by a high-temperature nebula and the vapor driven off by a strong solar wind. A third hypothesis is that after Mercury accreted and differentiated core from mantle, the planet was the target of a giant impact that stripped off and ejected much of the outer silicate fraction.

These three hypotheses, which differ strongly in their implications for how the inner planets came to differ in bulk composition, are testable because they predict different outcomes for the major-element chemistry of the silicate fraction of the planet. Mariner 10 carried no chemical remote sensing instruments, however, and ground-based efforts to deduce compositional information about Mercury's surface from the identification of mineral absorption features in reflected visible and infrared radiation have had only limited success. Sorting out how Mercury ended up a dominantly iron planet requires chemical remote sensing from an orbiting spacecraft.

Mercury's magnetic field

One of the major discoveries of Mariner 10 was that Mercury has an internal magnetic field. This was a surprising finding, because a planet as small as Mercury should have cooled over its lifetime to a greater extent than Earth. Earth's magnetic field is known to arise through the dynamo action of convective motions in its fluid metal core. Numerical models of interior cooling predict that a pure iron core in Mercury would have fully solidified by now. The field detected by Mariner 10 appears to be predominantly dipolar, like Earth's field, but the dipole moment is smaller by a factor of about 10^3.

An Earth-like hydromagnetic dynamo in a fluid outer core is only one of several ideas postulated to account for Mercury's magnetism. A fossil field in Mercury's crust remaining from an earlier era when a core dynamo was active is another possibility, and more exotic dynamos (e.g., thermoelectric currents driven by temperature variations at the top of a metal core with a bumpy outer boundary) have also been suggested. These hypotheses can be distinguished because they predict different geometries for the present planetary field and magnetic field measurements made from an orbiting spacecraft can separate internal and external fields and map the internal field.

Mercury's magnetosphere — the region of space in which the motion of charged particles is dominated by the planetary field — is the most similar to Earth's magnetosphere among the planets, but with important differences (Fig. 3). The interplanetary magnetic field of the solar wind is stronger closer to the Sun; Mercury occupies a much larger fractional volume of its magnetosphere because of its weaker internal field; and Mercury lacks an ionosphere, the site of closure of current systems in Earth's magnetosphere. Mercury's magnetosphere is therefore an important laboratory for generalizing our understanding of Earth's space environment.

Mercury's geological history

The geological history of Mercury has been deduced from the images taken by Mariner 10, but there are many unanswered questions. Mercury's surface consists primarily of heavily cratered and smooth terrains (Fig. 4) that are at least superficially similar in morphology and relative stratigraphic relationship to the highlands and geologically younger maria, respectively, on the Moon. Whereas the lunar maria are known to consist of basaltic lava flows on the basis of samples returned by the Apollo missions and orbital images of frozen lava flow fronts in several maria, the smooth plains on Mercury are higher in albedo (i.e., brighter in reflected light) than the lunar maria and no volcanic features can be seen in the relatively coarse-resolution Mariner 10 images. The role of volcanism in Mercury's history is therefore an open issue.

From the standpoint of large-scale deformation, Mercury shows evidence of an interesting history. The most prominent deformational features are the lobate scarps (Fig. 5), thought to be the surface expression of large thrust faults produced by horizontal shortening of the crust. The apparently random distribution of these scarps on all terrain types has led to the interpretation that they are the product of global contraction — a shrinkage of the planet as the interior cooled and the core solidified. Global shrinkage was at one time suggested as an explanation for the formation of mountain systems on Earth, but that idea was discarded with the acceptance of plate convergence at subduction zones. Mercury may

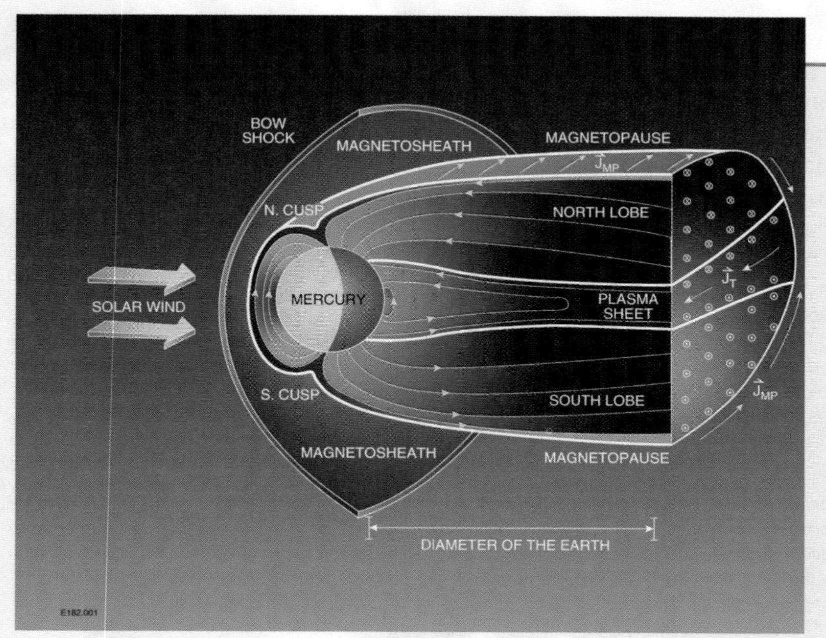

Left, Figure 3: *Observations by Mariner 10 and extrapolations from spacecraft measurements near the Earth suggest that the magnetosphere of Mercury is a miniature version of the Earth's magnetosphere generated by the interaction of the Earth's internal magnetic field with the solar wind. Many details of Mercury's magnetic field and magnetosphere are not understood, however, in large part because of the limited sampling by Mariner 10. (Figure courtesy James A. Slavin, NASA Goddard Space Flight Center.)*

Above, Figure 4: *Mariner 10 images of Mercury were obtained with three color filters. This mosaic, with false colors selected to emphasize spectral variations with chemistry and mineralogy seen on the Moon, illustrates that geological units on Mercury can be distinguished on the basis of color and information on mineralogy is derivable from surface spectral reflectance measurements. (Image courtesy Mark Robinson, Northwestern University.)*

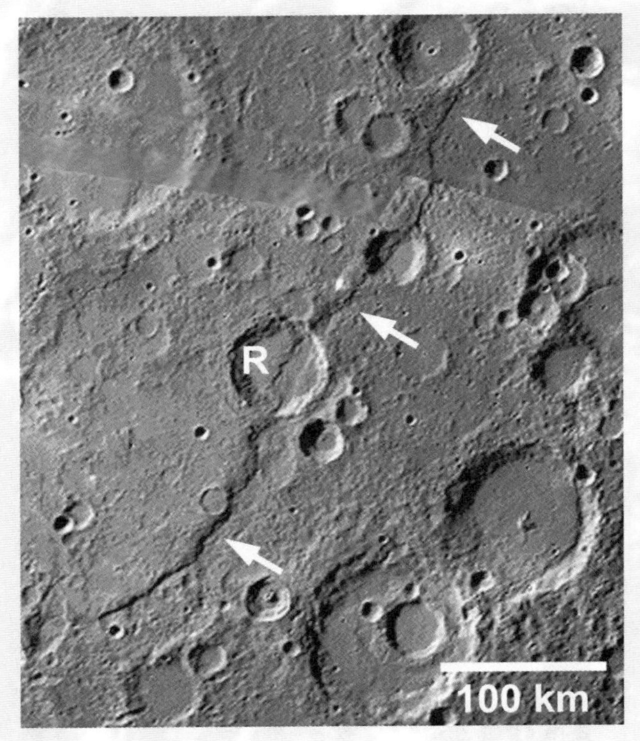

be the one planet where a record of such shrinkage is preserved. A critical test of that conclusion will be possible when images are taken of the hemisphere that Mariner 10 did not view.

Mercury's atmosphere and polar volatiles

Mariner 10 detected the presence of hydrogen, helium, and oxygen in Mercury's tenuous atmosphere, known as an exosphere. Ground-based spectroscopic observations led to the discovery of additional species, including sodium, potassium, and calcium. Most of these constituents are too abundant to be derived from the solar wind and their atmospheric lifetimes are much shorter than the age of the planet, so there must be steady sources at the planetary surface. The specific processes controlling the sources and sinks of atmospheric components are not well known. Key information from an orbiting spacecraft that would help to discriminate among competing hypotheses are the detection of additional species and the monitoring of exospheric properties as functions of time of day, solar distance, and level of solar activity.

One or more additional volatile species appear to be present at the surface near the planetary poles. Ground-based radar imaging of Mercury led to the discovery in 1991 of radar-bright polar deposits localized within the floors of near-polar impact craters (Fig. 6). The deposits have radar reflectivities and polarization characteristics that are well matched by water ice, although other materials have also been suggested. Ices are stable for billions of years in such areas because Mercury's obliquity (the tilt of its spin axis from the normal to the orbital plane) is nearly zero and the floors of near-polar craters are in permanent shadow and consequently very cold. Remote sensing measurements from an orbiting spacecraft are needed to confirm the composition of these trapped volatiles.

The long hiatus in Mercury exploration

Given the broad sweep of issues addressable with a Mercury orbiter, why did more than 30 years pass between the first Mariner 10 flyby of Mercury and the launch of the next mission to the innermost planet? The answer to this question has several parts. After Mariner 10's discoveries, there was widespread interest in a Mercury orbiter mission, but the change in velocity required for orbit insertion was thought to be too large for conventional propulsion systems.

In the mid 1980s multiple gravity-assist trajectories were discovered that could achieve

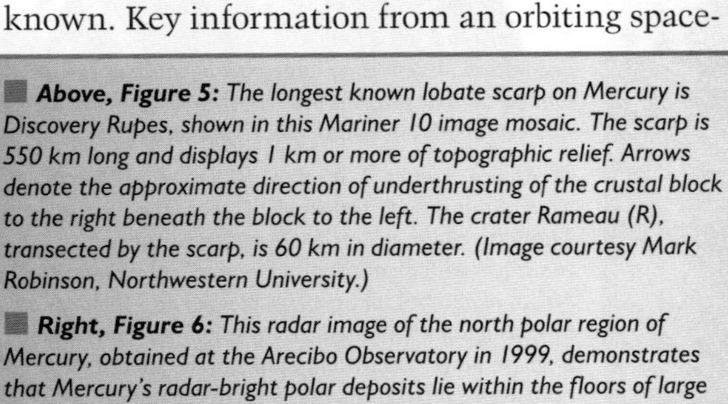

Above, Figure 5: *The longest known lobate scarp on Mercury is Discovery Rupes, shown in this Mariner 10 image mosaic. The scarp is 550 km long and displays 1 km or more of topographic relief. Arrows denote the approximate direction of underthrusting of the crustal block to the right beneath the block to the left. The crater Rameau (R), transected by the scarp, is 60 km in diameter. (Image courtesy Mark Robinson, Northwestern University.)*

Right, Figure 6: *This radar image of the north polar region of Mercury, obtained at the Arecibo Observatory in 1999, demonstrates that Mercury's radar-bright polar deposits lie within the floors of large impact craters. The radar direction is from the upper left, the resolution is 1.5 km, and the image is shown in polar projection. (Image courtesy John Harmon, National Astronomy and Ionosphere Center.)*

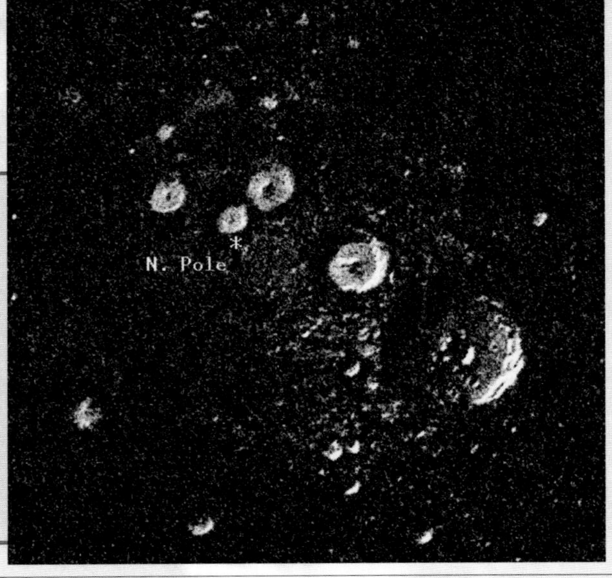

Mercury orbit insertion with existing propulsion systems. The 1980s were a difficult era in U.S. planetary exploration, however, marked by ambitious but costly and infrequent missions. By 1989, when the Galileo mission to Jupiter and the Magellan mission to Venus were launched, 11 years had passed since the previous U.S. planetary mission had left Earth.

In the early 1990s, after re-examining its approach to planetary exploration NASA initiated the Discovery Program, intended to foster more frequent launches of less costly, more focused missions selected on the basis of rigorous scientific and technical competition. Mercury was the target of a number of early unsuccessful proposals to the Discovery Program. The MESSENGER mission concept was born when engineers and space scientists at The Johns Hopkins University Applied Physics Laboratory (APL) came up with mission and spacecraft designs that looked practicable.

MESSENGER selection and development

In 1996, we assembled a team of scientific investigators and proposed a set of payload instruments that could make all of the global measurements discussed above. Our mission concept was awarded funds for further study, but it was not selected for flight, in large part because of concern with the ability of the spacecraft to survive the harsh thermal environment at Mercury. APL carried out an extensive testing of critical spacecraft components under high-temperature vacuum conditions, and we reproposed in 1998. After a thorough second review, MESSENGER was selected for flight in July 1999.

Between mission selection and launch were five event-filled years. We saw multiple changes in programmatic management at NASA and heightened concern for mission risk. The MESSENGER team faced myriad challenges — driven in large part by Mercury's unforgiving thermal environment and a stringent limit on spacecraft mass — including component and manufacturing issues, changes in personnel, and consequent schedule delays. A robust and thoroughly tested spacecraft nonetheless was delivered to Cape Canaveral, mated to its Delta II launch vehicle (Fig. 7) and successfully sent on its multi-year journey toward Mercury.

Above, Figure 7: *The complex process of assembling and testing the MESSENGER spacecraft and mating it to its launch vehicle extended over a year and a half. Shown is the spacecraft on July 14, 2004, after it was attached to the payload assist module of the Delta II third stage at Astrotech Space Operations in Titusville, Florida. The two flat, reflective panels are the solar arrays stowed in their launch positions. (Image courtesy NASA and The Johns Hopkins University Applied Physics Laboratory.)*

Left, Figure 8: *MESSENGER's wide-angle camera took this image of the Earth on 2 August 2005 shortly before closest approach during a gravity-assist maneuver. Portions of North, Central, and South America are visible. (Image courtesy NASA and The Johns Hopkins University Applied Physics Laboratory.)*

Onward to Mercury

En route to Mercury, MESSENGER will experience six planetary flybys, one of the Earth, two of Venus, and three of Mercury. The Earth flyby was completed successfully on 2 August 2005, and observations of both Earth and Moon provided important calibrations of most of the MESSENGER instruments (Fig. 8). The Venus flybys, in October 2006 and June 2007, will provide opportunities for both additional calibrations and new science. The first Mercury flyby, in January 2008, will permit close-up views of approximately half of the hemisphere of Mercury not seen by Mariner 10, and most of the remainder of that hemisphere will be viewed during the second and third Mercury flybys in October 2008 and September 2009. At its fourth encounter with Mercury, in March 2011, MESSENGER will fire its large thruster and enter into orbit about the planet closest to our Sun.

Thirty-six years to the month after the final Mariner 10 flyby of Mercury, the innermost planet will gain a new artificial satellite, a probe that will characterize its surface, interior, and environment on a global and continuing basis. Mercury will be neglected no longer.

Sean C. Solomon is the Principal Investigator for the MESSENGER mission to Mercury and the Director of the Department of Terrestrial Magnetism of the Carnegie Institution of Washington. A veteran of NASA's Magellan mission to Venus and the Mars Global Surveyor mission, he was a Professor at the Massachusetts Institute of Technology for more than 20 years.

Ralph L. McNutt, Jr., is the Project Scientist and a Co-Investigator for the MESSENGER mission to Mercury and a member of the Principal Professional Staff of the Applied Physics Laboratory of The Johns Hopkins University. He is also a science team member for the New Horizons, Cassini, and Voyager missions.

■ *Artist's impression of the MErcury Surface, Space ENvironment, GEochemistry, and Ranging (MESSENGER) spacecraft in orbit at Mercury. MESSENGER launched from Cape Canaveral Air Force Station on 3 August 2004, and will begin a yearlong orbital study of Mercury in March 2011. Image courtesy NASA/Johns Hopkins University Applied Physics Laboratory/Carnegie Institution of Washington*

4.2

Visiting VENUS...

At the dawn of the 'space age', about all that was known for certain concerning the planet Venus was that:

It takes 228 days to orbit the Sun

It is completely enshrouded in cloud

A faint Y-shaped pattern in ultraviolet light suggested the upper atmosphere circulates in about 4 days

A radio telescope serving as a radar revealed that the planet itself takes 243 terrestrial days to rotate and, astonishingly, its axis is inverted

A radio telescope acting as a microwave radiometer suggested the surface is very hot.

Here, *David Harland* takes a close look at Earth's cloudy neighbour.

EXPLORATION RESULTS from VENUS

IN 1961, the Soviet Union became the first to dispatch a spacecraft to study Venus. Although contact with Venera 1 was lost when it was some 7.5m km from Earth, the mission represented a partial success in that the 'escape stage' had set up a trajectory that took the probe within 100,000km of Venus that May, thereby demonstrating that interplanetary flybys were feasible. NASA's Mariner 2 made a 35,000km flyby on 14 December 1962, during which a microwave radiometer established the temperature near the surface to be 480°C.

Venera 2 fell silent shortly prior to a 24,000km flyby in February 1966. Although Venera 3 was able to refine its trajectory to 'hit' the planet the following month, by then it, too, had succumbed to some sort of systems failure. However, Venera 4 managed to survive the long interplanetary flight and an hour prior to its arrival on 18 October 1967 it released an entry probe. The trajectory and therefore the entry point were constrained by the requirement for the probe to penetrate the atmosphere at precisely the right angle, for it would burn up if this were too steep and bounce off if it were too shallow. It also required a clear line of sight to Earth in order to transmit its results.

Venera 4 penetrated the equatorial zone, in darkness, 1,500km beyond the terminator. Having slowed to 300m/sec, the 1m diameter probe deployed a parachute to slow its rate of descent and made *in situ* measurements until the ambient pressure was 22 bars. A nitrogen atmosphere with up to 10% carbon dioxide had been expected, but the concentration of carbon dioxide was revealed to exceed 90%, which went a long way towards explaining the extreme conditions. The Soviets said the probe had been disabled upon striking the surface, but this conflicted with the final temperature reading of 280°C.

When Mariner 5 flew past the planet the following day, its trajectory took it behind the planet as seen from Earth and measurements of the manner in which its signal was attenuated by the planet's atmosphere indicated the surface pressure to be in the range 75 to 100 bars, which meant that when Venera 4 fell silent it had still been 27km above the surface.

Inset: *The faint Y-shaped marking in the upper atmosphere of Venus is visible in this ultraviolet image obtained by the Pioneer Venus 1 orbiter spacecraft on 26 February 1979. Image courtesy NASA/JPL.*

Above: *On its mission to Venus, Mariner 2 became the second spacecraft to fly by another planet. During its journey, the craft for the first time measured the solar wind, a constant stream of electrically-charged particles flowing outward from the Sun.*

Striving to reach the surface

As an interim step, while a stronger probe was developed, it was decided to send one with a smaller parachute that would enable it to descend more rapidly and so penetrate deeper prior to succumbing. On 16 May 1969, Venera 5's final readings at an altitude of 25km were 320°C and 27 bars, from which it was inferred that the surface pressure was about 140 bars. This probe confirmed the high carbon dioxide content (95%) with most of the remainder being an inert gas (later established to be nitrogen); oxygen comprised at most 0.4%. The hot lower atmosphere was arid. The upper atmosphere was cooler and although it contained water vapour it was by no means saturated and the clouds were not water droplets. In an effort to reach the surface, the Soviets re-engineered their probe to withstand 180 bars.

After a parachute descent of 26mins on 15 December 1970, Venera 7 appeared to fall silent, but subsequent analysis found that the signal had continued at barely 1% of its previous strength for a further 23mins, from which it was concluded the probe had rolled over on landing and pointed its antenna away from Earth. It was also discovered that the telemetry system had malfunctioned by fixating on the temperature sensor. However, by taking into account how the probe had been slowed during its descent (as shown by the Doppler on its signal) it was possible to infer how the pressure varied with altitude and thus determine that the pressure at the surface was almost 100 bars. The pressure of the air on the surface of Venus was therefore comparable to that of water in a terrestrial ocean at a depth of 1,000m.

The next step was to trim the pressure shell to deal with 'only' 100 bars and to use the mass thus saved to improve the thermal shielding and thereby increase the operating life of the next probe. In addition, the antenna was modified to enable the probe to maintain contact no matter how it came to rest. Venera 8's arrival on 22 July 1972 was just after local dawn. It operated for 50mins on the surface and measured the temperature as 470 (+8)°C, the pressure as 90 (+2) bars and the composition as 96% carbon dioxide, 3% nitrogen and at most 0.1% oxygen. During its descent, the wind speed declined from 100m/sec at an altitude of 48km to less that 1m/sec at 10km, revealing the dense lower atmosphere to be stagnant. A gamma-ray spectrometer suggested a potassium-rich granitic rock. This success drew to a close the first phase of Soviet exploration.

When Mariner 10 flew by Venus in 1974, it was using the planet's gravity for a 'slingshot' to Mercury. This spacecraft had a camera. No detail was evident in the visible spectrum, even from the 5,750km point of closest approach, but an ultraviolet filter revealed the Y-shaped pattern, resolving structures in the air flow only a few kilometres across. This rapid four-day circulation prevents the atmosphere on the night side from cooling down; indeed, the infrared radiometer found the temperature of the cloud tops to be a uniform –23°C across both hemispheres.

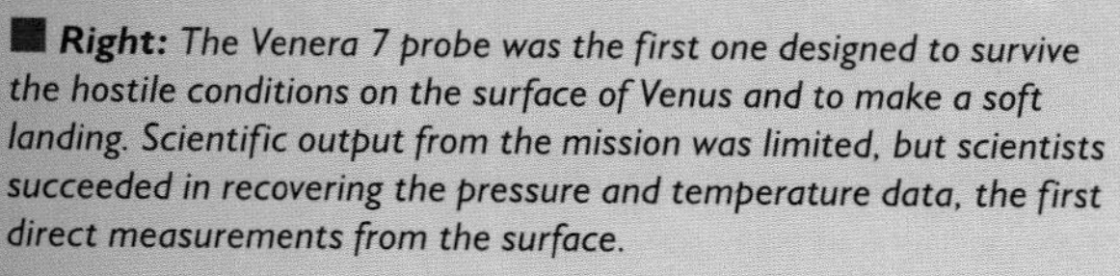

Right: *The Venera 7 probe was the first one designed to survive the hostile conditions on the surface of Venus and to make a soft landing. Scientific output from the mission was limited, but scientists succeeded in recovering the pressure and temperature data, the first direct measurements from the surface.*

Above: *Mariner 10 was the first spacecraft to use the gravitational pull of one planet (Venus) to reach another (Mercury), and the first spacecraft to visit two planets. Mariner 10 was also the first (and to date the only) spacecraft to visit Mercury. Mariner 10 returned the first-ever close-up images of Venus and Mercury.*

Inspecting the surface

Having demonstrated that their probes could survive the descent, the Soviets were eager to see what the surface looked like. The requirement to have a clear line of sight with Earth in order to enable the atmospheric probes to transmit to Earth had effectively restricted them to the night side (admittedly, Venera 8 had entered on the illuminated side of the terminator, but the Sun had been only a few degrees above the horizon). For a probe to land in full daylight, it would be necessary to provide a relay for its transmission. After releasing its probe about two days from the planet on a trajectory to intercept the trailing hemisphere, the new spacecraft would execute a deflection manoeuvre in order to pass 1,600km ahead of the planet, at which point it would enter an eccentric orbit. It would relay its probe's signal to Earth during the rise towards its initial apoapsis, then undertake remote sensing over an extended period. Following initial braking, the new probes were to deploy parachutes in order to provide time to sample the cloud layer as they descended from 65 to 50km, then jettison their chutes and pass through the inhospitable lower atmosphere as fast as possible and thereby maximize their surface time; nevertheless, the air was so dense that they struck the ground at a mere 5m per second. The spherical instrument unit was set on a ring-shaped shock absorber which was sufficiently wide to ensure the probe would remain upright even if it came to rest on a slope.

On landing on 22 October 1975, Venera 9 transmitted the first image of the planet's surface. This was a 180° monochrome panorama compiled over 20mins by a line-scan facsimile camera. The air was transparent, with 2.5% of the incident insolation penetrating the total overcast to illuminate the surface, but the light was so diffused that there was no hint of the Sun's position in the sky. The probe sat on a 20° slope littered with small angular rocks.

Three days later, Venera 10 landed 2,500km to the south, in a different landscape. In contrast to rocks sitting on the surface, there were fewer and rather slabbier rocks that were either a remarkably level outcrop or a fragmented crust laid down as a thin sheet of lava. The gamma-ray spectrometers indicated both sites to be basaltic in nature. Venera 11 and Venera 12 in December 1978 were disappointing. As a result of delays in their preparation they missed the optimum launch window and had to be launched without the propellant intended to enable them to enter orbit and hence were limited to flyby trajectories that would allow each departing craft to relay its probe's transmission to Earth. The improved probes had more instruments, both to study the atmosphere and the surface, but the atmospheric data was ambiguous and the surface imaging was frustrated when the lens caps failed to release!

When Venera 13 landed in March 1982, it had a pair of improved cameras which made colour panoramas that scanned the entire horizon. It landed on an exposure of slabby rock that was intensely fractured and laced by small lithic fragments and fines. The rocks appeared orangey, but this was because the blue and green parts of the spectrum were absorbed by the dense atmosphere, leaving just the red wavelengths to illuminate the surface; colour-calibrated images showed the rocks to be gray. For the first time, an X-ray fluorescence spectrometer was used to measure the chemical composition of the ground. Its results suggested a composition similar to a potassium-rich alkaline basalt. Venera 14's site, 1,000km to the southeast, was also a slabby outcrop with a layered structure that suggested the

■ ***Above:*** *Veneras 13 and 14 were identical Soviet spacecraft built to take advantage of the 1981 Venus launch opportunity and launched 5 days apart.*

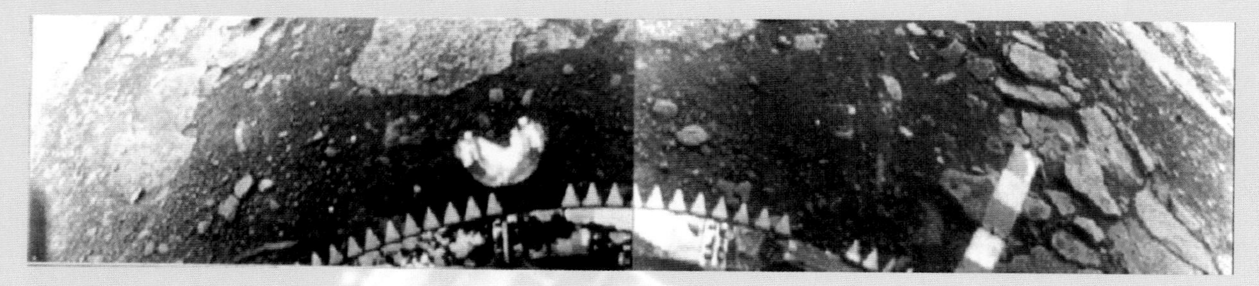

surface was a succession of thin sheets of lava. The less potassium-rich rock resembled tholeiitic basalt, a primary magma that is erupted at mid-ocean rifts on Earth.

The weather system

On 4 December 1978 NASA placed Pioneer Venus 1 into orbit for remote sensing of the planet's atmosphere and radar mapping of its surface. Pioneer Venus 2, trailing several days behind, was a bus carrying four atmospheric probes. The principal probe was released at a range of 11m km, on a trajectory to enter the equatorial zone on the dawn side of the terminator. When 1m km closer, springs scattered the three smaller probes to follow slightly diverging paths with arrivals a few minutes apart: one far to the north on the night side; a second just south of the equator in full daylight; and the third closer to the equator near local midnight. The large probe deployed a parachute at an altitude of 68km to slow its descent while it sampled the clouds, then shed this at 47km to pass more rapidly through the inhospitable lower atmosphere. The small probes did not have chutes. As the probes could not sample their environment during the initial deceleration, the spacecraft's bus had mass spectrometers to report on the condition of the upper atmosphere before it burned up. Although not expected to survive the impact, the 'day' probe transmitted from the surface for over an hour, ceasing only when its battery was exhausted.

The overall results showed there to be a layer of fine haze between 90km and 70km altitude, the main cloud deck extended down to 48km, the air was clear down to a thin layer of haze at 31km, below which it was clear down to the surface. The maximum opacity was at an altitude of 50km, just above the base of the main cloud deck. Because the hot, dense, lower atmosphere is stagnant, the base of the troposphere (i.e. the convective part of the atmosphere, which on Earth contacts the surface) resides at an altitude of 48km.

By October 1992, when Pioneer Venus 1 finally fell into the atmosphere and burned up, it had far exceeded its design life and its data had been used to 'write the book' on the atmosphere and its interaction with the solar wind. The overall circulation system comprises a single Hadley 'cell' on each side of the equator: the air rises in the equatorial zone, flows at high altitude into the polar zone and then descends. Because the returning air flow is in the middle atmosphere, the surface is isolated from the 'weather system'. This simple circulation system derives from the fact that the planet rotates so slowly. On the rapidly rotating Earth, the Coriolis effect disrupts this simple circulation by inducing swirling air flows and tropical, temperate and polar components. In Venus's upper atmosphere, the winds of 100m/sec race from east to west, with the planet's retrograde rotation and the Y-shaped pattern derives from the fact that the poleward motion is at barely 10% of this rate. In fact, the weather system is more complex than this, as there is turbulence where sunlight penetrates deeper into the atmosphere at the subsolar point, forming vigorous isolated convection cells that cause hot gas to 'bubble out' of the top of the cloud deck and turbulent eddies as the zonal flow passes through this area on its race around the planet.

Orbital mapping

The radar on Pioneer Venus 1 had a surface resolution of only 75km, similar to that of the early terrestrial studies, but as an orbiter it had the advantage of being able to chart the entire globe at the same resolution and thereby provide a sense of perspective. Having sampled the surface at several locations, the Soviets decided to improve on the Pioneer map. The bus of the Venera spacecraft was modified to deploy a radar antenna and fitted with a second pair of solar panels to provide the power to operate the instrument. Two of these mappers entered orbit in October 1983 as Venera 15 and Venera 16. They assumed highly elliptical polar orbits with periapses at latitude

■ ***Above:*** *Venera 13 landed on Venus in 1982, returning the first colour images from the Venusian surface. Along with its twin, Venera 14, it sent back the first X-ray spectra from the surface of Venus, giving important compositional data. Image courtesy NASA.*

60° north, which enabled them to map as they dipped down to their 1,000km periapsis, then transmit the data and recharge their batteries on the rise to apoapsis. Each mapping pass imaged a strip that was 150km in width and 7,000km in length, running from 30° north latitude across the pole. As the planet slowly rotated beneath the planes of their orbits, they extended their coverage to the full range of longitudes.

The resulting 2km resolution map yielded some significant insights. In particular, the impact cratering suggested that the low-lying plains, which comprise the majority of the surface, were formed recently – perhaps just 500m years ago. There were tantalizing hints that the lithosphere undergoes global plate tectonics, but the evidence was weak. A better map was required. NASA's Magellan spacecraft entered an elliptical polar orbit of the planet on 10 August 1990 and, during the next four 243-day cycles, mapped 99% of its surface at a resolution of 120m, confirming it to be an intensely volcanic world, but one controlled by a style of tectonism which does not occur on Earth.

A runaway planet

In 1969, based on measurements of the index of refractivity, it was suggested that the condensates in Venus's atmosphere were acid-laden water droplets. The large Pioneer probe confirmed 75% to 85% of the cloud to be composed of sulphuric acid aerosols. These are believed to form by photochemical oxidation at altitudes exceeding 60km – the dissociation of carbon dioxide or sulphur dioxide in this zone yields atomic oxygen, which oxidizes SO_2 to SO_3, which is hydrated into sulphuric acid (H_2SO_4) droplets. As the droplets fall, they are thermally disrupted on reaching the 100°C temperature zone at an altitude of 49km, yielding SO_3 which, upon encountering carbon monoxide, regenerates sulphur dioxide and carbon dioxide. This precipitation cycle operates in the upper atmosphere, where the temperature is moderate. There is a steep thermal profile, with the temperature decreasing by about 8°C per kilometre of elevation. Hence, although for a specific elevation the temperature is uniform both from pole to pole and from daylight into the darkness, the surface temperature varies across the 13km range from the summit of the highest mountain to the floor of the deepest depression. At no point on the surface is water stable. The planet's remaining water is in the cooler upper atmosphere, but is progressively diminishing because the hydrogen atoms released by photodissociation will tend to escape to space. The liberated hydroxyl radicals will readily react with sulphur dioxide and enhance the production of sulphuric acid. Any free oxygen that reaches the surface will oxidize the hot rock, removing it from the atmosphere. Sulphurous gases are present in our own atmosphere only in trace amounts. They

are released by volcanic activity, but unless this is on a vast scale the sulphur is either dissolved by water droplets in clouds or, in arid regions, is bound up with oxygen-rich radicals or other trace gases to form particulates that fall to the ground. When an explosive volcano projects a sulphurous plume into the stratosphere, the aerosols of sulphuric acid that form are rapidly distributed around the globe, where, by reflecting sunlight, they chill the lower atmosphere.

Although most of the sunlight reaching Venus is reflected to space by the aerosol clouds, the 'greenhouse effect' of the high concentration of carbon dioxide is able to maintain the lower atmosphere at a high temperature. One theory is that the volcanism that almost totally resurfaced the planet 'pumped up' the atmosphere with sulphurous chemicals and triggered a 'runaway' process. As the hot lower atmosphere is arid, sulphur dioxide emitted by volcanic eruptions

■ **Above:** *Venera 15 was one of two Soviet spacecraft (along with Venera 16) designed to use 8cm band side-looking radar mappers to study the surface properties of Venus. Together, the two spacecraft imaged the area from the north pole down to about 30 degrees north latitude over the 8 months of mapping operations.*

cannot be washed out: once released it will remain in the atmosphere and provide a reservoir for the chemical reactions high above that produce the sulphuric acid aerosol clouds.

Ballooning

The Soviet Union's final missions to Venus were something of an afterthought: a pair of spacecraft were dispatched to rendezvous with Halley's comet and in June 1985, while using a Venus flyby to 'slingshot' them on to the comet, they released probes. In view of their dual objectives, these spacecraft were named VeGa (because the Russian alphabet does not include an 'H' (for Halley), 'G' was used instead and the spacecraft were named Venera–Gallei, which was contracted to VeGa). In a novel addition to the usual mission, as they penetrated the main deck of cloud each lander ejected an instrument package that inflated a helium balloon. At an altitude of 60km, the pressure was just 0.5 bar and the temperature was a moderate 40°C. An international network of antennae provided continuous monitoring of the transmissions from the balloons and the Doppler effect enabled them to be tracked drifting with the prevailing wind. Each balloon's battery could sustain two days of operation, during which the zonal flow would carry it half way around the planet and then out of contact.

■ **Above:** *An artist's impression of the surface of Venus. Image courtesy ESA.*

They were released at local midnight, four days apart, the first just north of the equator and the second just south of it. Both drifted some 11,500km over the terminator, into daylight. Whereas the first had a fairly smooth passage, the second, which passed over terrain which the radar map indicated to be elevated, had a wilder ride and at one point was drawn down several kilometres by a strong down-draft. Because the balloon missions obliged the probes to land in the dark, this time they did not carry cameras. However they had drills to recover samples for their X-ray spectrometers. Unfortunately, as the first lander headed for a low-lying plain its drill deployed prematurely. The second landed in the elevated terrain, functioned properly and found an alumina-rich rock chemistry suggestive of ancient crust, which was consistent with the topography.

A new window

In 1988, it was realized the upper atmosphere was transparent to narrow spectral bands in the near-infrared, providing 'windows' to observe the middle atmosphere. In principle, the absorption features at wavelengths in these bands would allow the minority constituents of the atmosphere to be determined all the way down to the surface. On its indirect route to Jupiter, NASA's Galileo spacecraft flew by Venus on 10 February 1990. Its infrared spectrometer was able to make such 'soundings'. At a wavelength of 2.3 microns

it was able to observe the turbulence at the base of the cloud deck. The radiant heat from the surface was attenuated by the sulphuric acid clouds and the spatial resolution was a poor 50km, but the different terrains (as identified by radar) were evident because the lower-lying terrains were hotter. The surface detail could be enhanced by subtracting the 2.3-micron signal representing the upper atmosphere from the 1.18-micron signal from the hot lower atmosphere and surface – a process dubbed 'de-clouding'.

Galileo's plasma wave spectrometer detected broadband electromagnetic pulses from lightning in the lower atmosphere. On Earth, lightning occurs when updrafts cause moist air to condense out droplets of rain and the rapid air motions promote the build-up of static electricity that discharges between clouds and to the ground; but Venus's lower atmosphere is arid. Terrestrial lightning can also occur in the plume of gas and dust emitted by an explosive volcano, because the particles in the roiling cloud are charged. The tremendous pressure at Venusian the surface would inhibit plume activity, but might its lightning be due to volcanic activity? Remote sensing by Pioneer Venus 1 showed that between 1978 and 1986 the amount of sulphur dioxide in the atmosphere at an altitude of 80km diminished by a factor of ten, prompting the proposal that a major eruption had occurred shortly prior to the spacecraft's arrival. However, when the Cassini spacecraft made a very close flyby on 26 April 1998, its radio and plasma wave spectrometer detected no evidence of lightning, suggesting that if the production of lightning was related to volcanism, this activity was not continuous.

■ ***Top:*** *This image of Venus was acquired by the Galileo spacecraft on 14 February 1990. A high-pass spatial filter has been applied to emphasize the smaller scale cloud features, and the rendition has been colorized to emphasize the subtle contrasts in the cloud markings and to indicate that it was taken through a violet filter. Image courtesy NASA/JPL.*

■ ***Above:*** *The launch of Starsem flight ST14 and a Soyuz FG-Fregat rocket carrying Venus Express, ESA's first probe to Venus. The mission successfully launched from Baikonour launch pad number 6 on 9 November 2005. Image courtesy ESA.*

Venus Express

The European Space Agency launched Venus Express on 9 November 2005, on a trajectory to arrive at Venus on 11 April 2006. It is to use sensors operating across a wide range of wavelengths, including the windows in the near-infrared, to study the composition and circulation of the atmosphere and how this interacts with the solar wind, and to seek chemical concentrations and 'hot spots' on the surface that would confirm that there is ongoing volcanic activity.

Dr David M Harland *gained his BSc in astronomy in 1977 and a doctorate in computational science. He has subsequently taught computer science, worked in industry and managed academic research. In 1995 he retired and has since published many well known books on space themes.*

■ ***Above:*** Artist's impression of Venus Express' main engine burn during Venus Orbit Insertion (VOI). The VOI manoeuvre took place successfully on 11 April 2006. Image courtesy ESA.

■ *An artist's impression of the European Space Agency's Venus Express spacecraft in orbit around Venus. After a 153-day cruise from Earth to Venus, the spacecraft entered Venusian orbit on 11 April 2006. After a series of orbital control manoeuvres, the spacecraft entered its nominal science operations orbit on 7 May 2006. Image courtesy ESA.*

4.3 EUROPA'S Ice next . . .

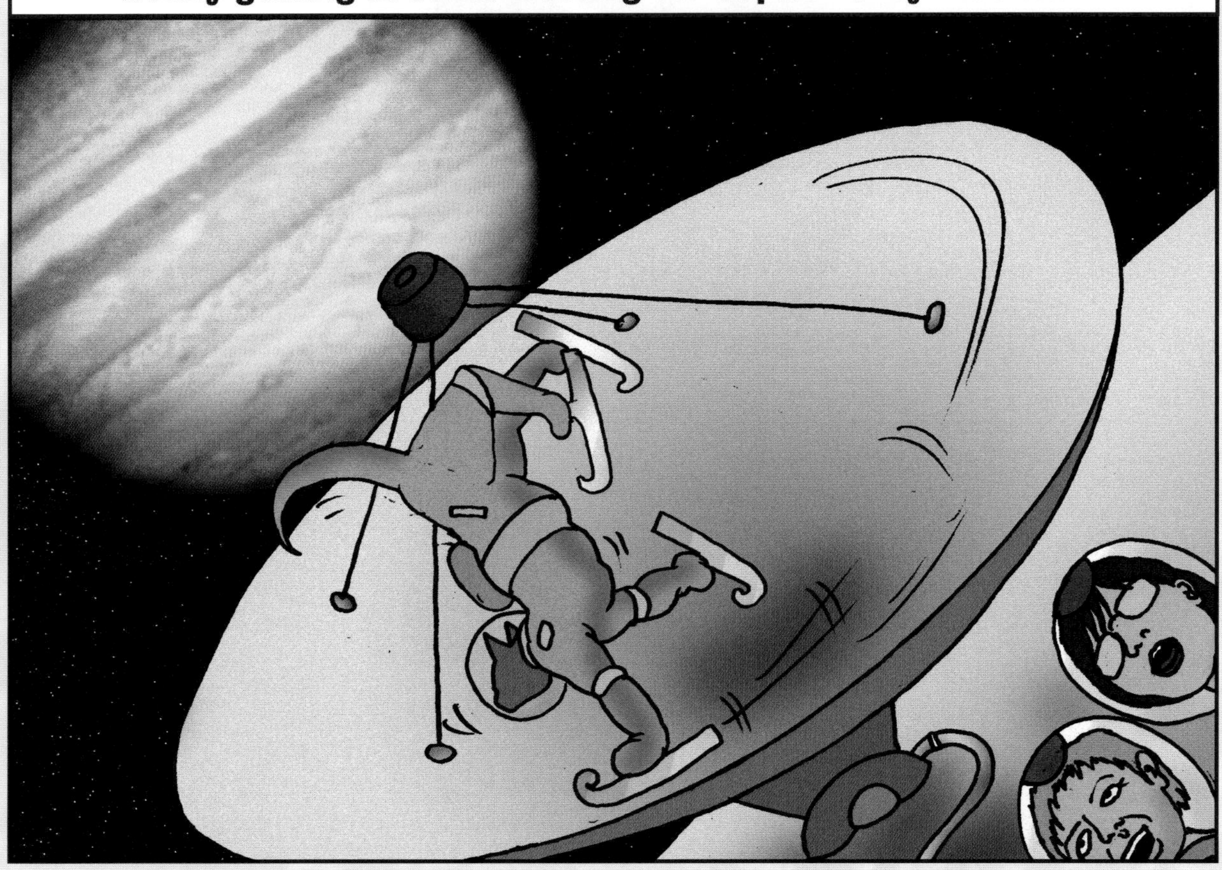

Jupiter's four large moons were first spotted by Galileo in 1610, a discovery that put to rest the notion that everything in the universe revolved around the Earth. As *Richard Greenberg* explains, three of those Galilean satellites contain large fractions of water, including surface ice that dominates their appearance. Inside, each also has significant rock and iron which play important roles: Europa's interior is mostly rocky underneath the H2O; Ganymede has a magnetic field, which means it must have a molten iron core; and Callisto probably contains a large fraction of rock mixed in, even near its surface. While the H_2O on each surface is frozen, inside each of these three "icy" moons lies a layer of liquid water.

Exploring JUPITER'S Icy Moons

A DEEP, LIQUID-WATER ocean lies just below the surface of Europa, so the external appearance has been continually renewed by shifting, melting, and refreezing of the ice . The surface is so young that, unlike our Moon, there are few craters. In fact the lack of craters means that the surface must have been entirely replaced during the past 50m years, which is only about 1% of the age of the solar system. At this rate, Europa's surface has been replaced almost twice-over in just the time since dinosaurs became extinct on Earth.

On Ganymede, the next farthest out from Jupiter, much of the surface has been modified by cracking and stretching, with some fluid flow,

Below: *(Figure 1) A color portrait of the icy satellites of Jupiter, with their sizes (Table 1) shown to scale. At the right is Callisto, the farthest from Jupiter, with its dark veneer. This surface is dominated by impact craters. To the left of Callisto, at the center of the montage, is Ganymede, with 1/3 of its surface similar to Callisto's, and 2/3 consisting of younger, bright grooved terrain. To the left of Ganymede is Europa, with global scale lines from tectonic cracking, and dark splotches marking thermally modified terrain. Craters are few on Europa, indicating that this surface has continually been reprocessed and may still be active. One crater, Pwyll, has a large system of bright ejecta rays, evident in this image. The tiny satellite Amalthea at the far left is the closest to Jupiter and the smallest of the group. Its elongated shape (Table 1)is barely visible at this scale. Amalthea was only recently recognized to be an icy body, a discovery that raises new questions about the origin of the Jupiter system.*

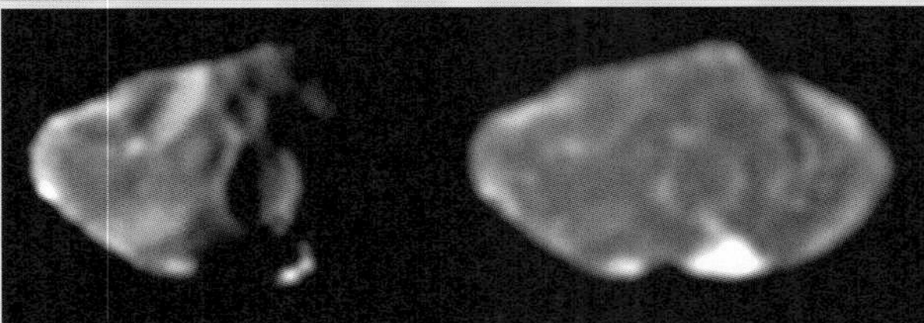

Left: *Two views of Amalthea, which has recently been found to have a low density indicative of a substantially icy composition. The bright crater at the south (bottom) may represent excavation to purer ice, similar to the bright craters on Callisto or on the dark terrain of Ganymede.*

Satellite	Discovery Year (1000 km)	Semi-major axis a	Radius (km)	Density (water=1)	Moment of inertia C/MR^2	H_2O (ice + liquid)	Liquid Ocean	Rock	Iron (+ FeS)
Europa	1610	671	1565±8	2.99±0.05	0.346±0.005	outer ~ 150km	Most of the H_20	Mantle below H_20	Core radius is 200-700 km
Ganymede	1610	1070	2631±2	1.942±0.005	0.312±0.003	outer ~ 900km	Layer below ~ 100 km ice	Mantle below H_20	Core radius is 650-900 km
Callisto	1610	1883	2410±2	1.834±0.003	0.355±0.004	Mixed with variable amount of rock to depth of ~ 1000 km	Layer below ~ 100 km ice	Mixed with ice in outer ~ 1000 km, mixed with iron below	Mixed with rock inside radius ~ 1200 km
Amalthea	1882	181	125x73x64	0.86±0.1		large fraction		Mixed with ice?	

Table 1

but the rest of the surface shows few signs of geological activity, other than bombardment by small asteroids and comets. Counts of the craters that record those impacts show that this part of the surface is at least 2bn years old, and maybe twice that age.

Callisto, the farthest of the Galilean satellites from Jupiter, is the most primitive of the three icy satellites. Its primordial composition has not been separated by gravity into layers, and its surface is dominated by ancient impact craters. Both in its orbital location and its physical properties, Callisto is the opposite extreme from the well-cooked, innermost Galilean satellite, Io, which is covered by volcanoes and devoid of water or ice.

The diversity among the Galilean satellites, especially in the varying fractions of water, results from their formation at different places and times within a cloud of dust and gas that surrounded Jupiter as that planet first formed. Later, the satellites changed in various ways as a result of two interrelated effects: (1) tides and (2) a resonance among the orbital periods of Io, Europa, and Ganymede. The internal structures and surfaces of those three moons were affected in various ways, while Callisto changed hardly at all. This chapter describes how these processes have played out differently on each of the icy moons, leading to the distinctive character of each one.

The story recently became more complicated by the discovery that Europa, Ganymede and Callisto are not the only icy moons of Jupiter. Jupiter has dozens of small moons in addition to the four large Galileans. Based on what scientists thought they understood about the formation of this system, they expected them to be rocky. Now, the discovery that Amalthea (Table 1) may be the iciest of all Jovian satellites is raising fundamental questions about the entire Jupiter system. Thus while this chapter describes current understanding of the satellites, continuing research ensures that the story will be evolving over the coming years.

Several spacecraft have contributed to the discoveries about the icy moons. First, Pioneers 10 and 11 flew by in 1973 and 1974. They took low-resolution images of the Galilean satellites, showing the moons for the first time as something more than astronomical points of light. Pioneer 11 passed close enough to Callisto to measure its mass. Pioneer 10 also discovered the intense radiation belts of Jupiter's magnetosphere, which are very important for the surfaces of the icy satellites.

The spacecraft Voyager 1 and 2 flew by in 1979. They obtained much better images of the satellites and improved measurements of the magnetosphere. The images showed the differences among the four Galilean satellites and demonstrated the crucial effects of tidal heating.

The Galileo spacecraft went into orbit around Jupiter in 1995 and actively took data there for eight years, flying close by all the Galilean satellites numerous times. Its data effectively superseded all the information from the previous spacecraft. Despite many technical difficulties, Galileo obtained many high resolution images of all the major satellites, determined their gravitational fields, and measured the magnetic effects of each satellite. Most of what we know about the icy satellites of Jupiter comes from Galileo spacecraft data.

Internal composition of Jupiter's icy moons

Information about the structure and composition of these satellites (Table 1) comes from several sources. The moons' gravitational effects on one another tells how massive they are, and how the mass is distributed inside them (from the *moment of inertia*). Images taken by spacecraft show the sizes of the moons, which combined with their masses tells us their densities and thus the possible types of materials inside. Our understanding of the substances likely to have condensed from the gaseous cloud that surrounded the early Jupiter tells us what types of materials are plausible. The magnetic fields near the moons have been measured by spacecraft instruments. From those data, we can infer the conditions of satellites' iron cores that generate magnetic fields and of satellites electrically conducting layers (probably salty liquid water) that modify Jupiter's magnetic field.

Putting all this information together gives us a good idea of the bulk internal properties and structure of each of the icy Galilean satellites.

Europa

Europa, with a radius of 1,565km, has a density of 2.99±0.05 gm/cm^3, much greater than water (density 1 gm/cm^3, by definition) or ice which is a few percent less dense than liquid water. From its measured moment of inertia (Table 1), we have information on the mass distribution. Internal heating by radioactivity and tidal friction must have allowed the less dense materials to float up toward the surface, so there is an outer layer of H_2O about 150km thick.

Tides are very strong on Europa (as described below). They probably generate enough frictional heat to keep most of this water in the liquid state, with only a thin shell of ice on the surface. Evidence that most of the water is liquid came from interpretation of characteristic crack patterns in the ice, especially the long chains of arcs, called "cycloids" (Figure 2). These crack patterns fit the stress expected from the tidal movements of the underlying ocean. Evidence that the ocean exists now, or at least up to the end of the 20th century, came from variations in Jupiter's magnetic field that could be explained by a salty ocean within 200km of the surface.

In fact the ocean comes quite close to the surface, as indicated by a variety of geological features (see below), which are best explained by linkage to the ocean below. Temporary openings in the ice shell by cracking and local warming suggest the ice is thinner than about 10km, but with thickness varying with time and place. The amount of internal tidal heating and the heat transport through the ice have frustrated geophysical modeling of the ice thickness. However, it is generally agreed that the bulk of the H_2O on Europa probably forms a deep liquid water ocean just below the ice.

Below the ocean lies the rocky "mantle" and the iron core. The radius of the core is between 200km and 700km, depending on assumptions about the amount of iron sulfide. Given the

■ ***Above:*** *(Figure 2) Cycloidal crack patterns (chains of arc), are ubiquitous on Europa, usually displayed in the form of double ridges lining the sides of the cracks. Each arcs is typically about 100km long, and many cycloids comprise a dozen arcs or more.*

tidal heating implied by the liquid water ocean, the core should be at least partly melted, a requirement for generating a magnetic field. However, the Galileo spacecraft found no internally generated magnetic field, suggesting either a solid core or that the liquid is not flowing. In bulk structure, Europa is only partly an icy satellite. But in terms of surface appearance it is dominated by water ice.

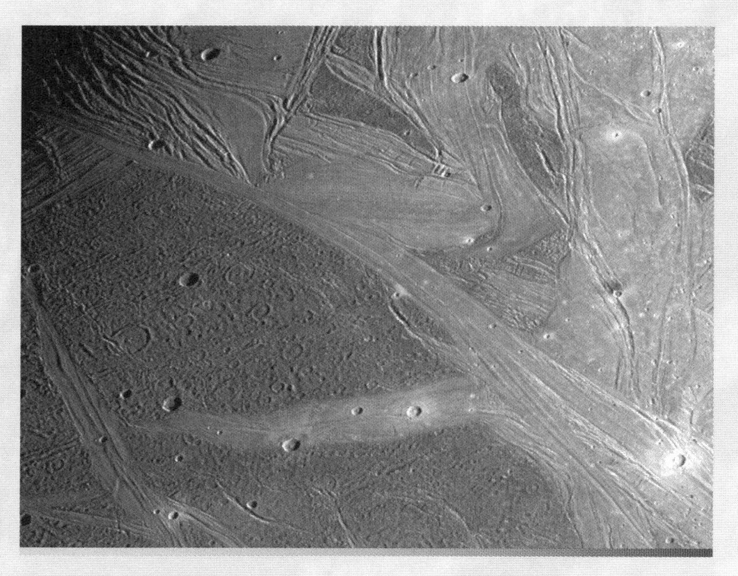

Ganymede

Ganymede's moment of inertia value (Table 1) is also low, implying (as for Europa) interior separation into layers of water, rock and iron. Of the total radius of 2,410km, the outer layer of H_2O is about 900km thick, the interior core is 650-900km in radius, and the rocky mantle lies in between. The heating that allowed this layering was provided early on by impact energy as small bodies were accreted by the growing moon and by radioactivity. Tidal heating due to the orbital resonance with Europa and Io may have contributed as well.

Ganymede does have a magnetic field, with a strength at the surface of about 1/10 of that of the Earth. To generate this field, the core must be molten. The Galileo spacecraft's magnetometer also detected variations in Jupiter's magnetic field that suggest a layer of liquid salty water. But in contrast to Europa, the conducting layer must be well below the surface. Accordingly, the surface geology shows no signs of interaction with the liquid layer.

Theoretical models of ice at great depth are consistent with the possibility of a liquid layer. The melting temperature of water decreases slightly with modest pressure (explaining for example why ice skates slide so easily), but as the pressure increases further, the melting point rises. For Ganymede, the minimum melting temperature would be at 150km below the surface, assuming pure water. Thus any liquid would be around that depth, with ice layers above and below it.

Estimates of the thickness of the outer layer of ice depend on the amount of heat generated inside and the efficiency of transporting that heat outward. For a reasonable heat flux and thermal conduction, the ice above the liquid layer would be about 100km thick. In that case, the liquid layer would also be thick, extending from depth 100km down past the minimum-melting-point depth of 150km. On the other hand, these values change if the outer layer of ice is transporting heat convectively, that is, by the viscous flow of ice. It that case, it conveys heat more efficiently so its equilibrium thickness is greater, leaving space for only a few kilometres of liquid at most. Salt in the water changes the story only slightly. However, ammonia, which is another plausible substance, could make a very big difference: An ammonia-water layer could be 200-300km thick and as close to the surface as 70km. All these estimates are very uncertain, but they show that a fluid layer is a plausible explanation for the measured magnetic effects.

Even though any liquid layer is probably too deep to have a direct effect, the surface of Ganymede experienced substantial geological activity and variability. The exception is the dark and heavily cratered terrain covering about one-third of the surface, with little internally driven geology. The large number of craters indicates that the surface is more than 4bn years old. The darkening may result from impurities such as fine bits of rocky material (silicates) that remained in the cold icy outer crust, even while the interior underwent the heat-driven differentiation that sent most of rock into a deeper interior layer. Sublimation of ice off the surface may have further concentrated dark material in a thin layer. Ice can be darkened considerably by such processes, because even a relatively small concentration of dispersed dark material will absorb a large amount of light.

Above: *(Figure 3) The bright grooved terrain on Ganymede contrasts with the dark ancient terrain in this much higher resolution view than that shown in Figure 1. The greater crater density on the dark terrain is evident, as is the predominantly tectonic character of the brighter terrain. This region is about 660 by 520km in size with resolution about 1km per pixel (JPL/NASA image).*

The remaining two-thirds of the surface has been geologically active, yielding brighter terrain with complex sets of roughly parallel grooves. The surface here seems to have been stretched and cracked, exposing brighter ice from below. In addition, this modified terrain shows evidence of volcanic resurfacing, a process often called *cryovolcanism*, because on Ganymede liquid water flows out and over ice, rather than molten lava flowing over rock.

The crater record on the bright two-thirds of the surface confirms that it is younger than the old, dark, heavily cratered one-third, perhaps about half as old. Evidently, global-scale internally driven processes occurred long after formation of the satellite, but then turned off very long ago. However, the crater-based age estimate is so uncertain, that the resurfacing could have been associated with the initial formation of the satellite, rather than a subsequent heating event. The range of uncertainty also admits the possibility that the surface is relatively recent. However, in contrast to Europa's surface which may still be undergoing continual renewal, Ganymede's has probably been relatively unchanged for at least 1/2 bn years and perhaps for as long as 4bn years.

Callisto

Callisto is of similar size and density to Ganymede (Table 1, Figure 1) but its moment of inertia is much larger. Thus, while its composition of ice, rock and iron is probably similar, there has been only partial differentiation into layers (note that fully-differentiated Europa has a moment of inertia closer to Callisto's than to Ganymede's because it has a relatively thin H_2O layer, so it is more uniform than Ganymede but for a different reason than Callisto is.) If Callisto were completely uniform, its moment of inertia would have coefficient 0.4; gravitational compression without differentiation would reduce that value to about 0.38. The actual value of 0.355 indicates that there has been at least some differentiation, but not much.

Most likely the partial differentiation has resulted in a distinctly ice-rich layer. The argument for that structure, rather than a fairly continuous increase in the fraction of rock and metal with depth, is the following. The ice layer would undergo convection, which rapidly transports out the internal heat, shutting off differentiation. Without the distinct ice-rich layer, convection would be suppressed, internal heat would have built up, and differentiation would have proceeded. Evidently, this scenario did not occur, so there must be some layering. However, there is a wide range of uncertainty about the thickness of the ice-rich layer, its purity, and the compositional profile below it.

While iron must be concentrated somewhat toward the center, there is no evidence for an iron core from either the gravitational field or magnetic field measurements. Similar to Ganymede however, the magnetometer data does suggest a deep conducting layer, again most plausibly explained by a liquid water layer. The depth, thickness, and composition of a liquid layer would be subject to the same constraints as discussed above for Ganymede.

Like Europa and Ganymede, Callisto's surface is dominated by water ice. The surface is darkened by the presence of clays (hydrated silicates) and organic chemicals (that is, complex compounds rich in carbon) that may have rained onto the surface from comets and asteroids, and likely concentrated by sublimation of the ice at the surface. Many craters, which dominate the geology of Callisto, appear as bright spots on the generally darker surface (Figures 1 & 4). The large number of craters shows the surface to be at least 4 billion years old, although some craters are younger.

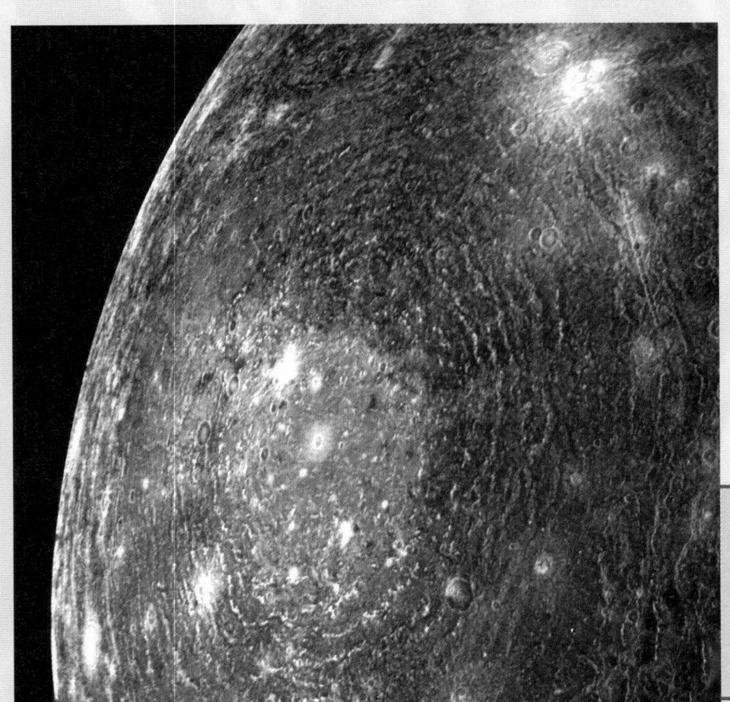

Above: *(Figure 4) Callisto's surface is ancient and heavily cratered (c.f. Figure 1). This image shows the huge multi-ringed impact feature Valhalla, with its 300-km-wide bright center, where underlying ice has been exposed. To the right, a string of small craters (a catena) is probably the result of impact with a comet that had been pulled apart by Jupiter's tidal effect, before hitting Callisto. Several such catenae are found on Ganymede and Callisto. (JPL/NASA image)*

Amalthea

The trend of compositions among the Galilean satellites, from rocky Io closest to Jupiter out to the icy Ganymede and Callisto is strong and is broadly consistent with expectations for formation in a cloud of dust and gas around the hot planet. Hence, Amalthea, which was discovered by Barnard in 1892 on a regular, circular orbit interior to the Galilean satellites was long assumed to have formed along with them and to be rocky. Its fairly dark surface fit that notion.

The Galileo spacecraft passed close enough to Amalthea (Figure 1b) to experience a slight gravitational pull by the satellite's mass. Combining the mass value with determination of its volume from images yielded a density of 1.0± 0.5 gm/cm^3. Because the density of most rock is closer to 3 gm/cm^3, this result was taken as evidence that the body must be quite porous, as a jumble of gravitationally bound rocky debris. Recently, however, analysis of the spacecraft trajectory has reduced the density value to 0.86 ± 0.1 gm/cm^3, a very low value that implies a substantial, probably dominant, ice component, and still requiring considerable porosity.

The presence of this icy satellite so close to Jupiter and in a very regular orbit means that models of the formation of the Jupiter system will need to be reevaluated carefully.

Formation and evolution of the satellites

Formation of the Jovian system

The characteristics of the icy satellites result from how they formed, modified by evolution dominated by tidal effects. Most important has been tidal heating, driven by the orbital resonance. In fact, the observed characteristics of the satellites place some of the strongest constraints on theoretical models of the formation and evolution of the Jovian system.

Jupiter itself formed from the gravitational collapse (probably around a core of solid material) of a clump of gas in the primordial nebula around the Sun. The newly formed planet was itself then likely surrounded by a flattened, spinning nebula of dust and gas. This nebula may have been left there during planet formation because it was rotating too fast to collapse, or else it formed from material entering the region around the planet from the circumsolar nebula. Formation of the regular satellites of Jupiter, including the Galilean satellites and Amalthea, has long been assumed to have progressed within the circumJovian nebula. Small particles of material condensed and then collided, accreting into ever larger bodies. In that sense, the Jupiter system was like a miniature solar system, with satellites forming the same way most planets formed in orbits around the Sun.

The Jovian nebula probably cooled enough for H_2O to condense only outside the orbit of Europa, explaining why Ganymede and Callisto accreted largely from water ice, while Europa contained relatively little. That model assumes that the Jovian nebula had (a) the same composition as the solar nebula and (b) the lowest possible mass that would contain enough condensable materials to produce the satellites.

That model had major shortcomings. First, as satellites formed in such a gassy medium, they would have rapidly spiraled in toward Jupiter long before the nebula dissipated. Second, the model did not explain why Callisto is only partially differentiated, because the rapid accretion would have resulted in enough heating to differentiate *all* of the Galilean satellites. Only slow accretion would have allowed Callisto to avoid complete internal differentiation, because the heat generated during each impact would need to radiate away before being buried by the accretion of more material.

Two theories have been proposed to avoid the rapid shrinking of the orbits and to account for the slow accretion of Callisto. One idea is that Callisto accreted much further out in the Jovian system where (with less material) accretion would be slower, and then it spiraled in to near its current orbit. In that model, the other satellites open gaps in the disk, a process that slows their orbital evolution. The other idea assumes a Jovian nebula that at no given time contained even near enough mass to build the satellites. Instead, this nebula was continually fed by the gas and dust cloud around the Sun on its way to accretion by Jupiter. The lower mass of the nebula at any given time slows both the growth and the orbital drift of the satellites, avoiding the problems with the more massive nebula. An open question is how Amalthea, on an orbit interior to Io's, was able to form with an icy composition.

Tidal effects of the Laplace resonance

Heating due to radioactivity and accretion were probably important during the formation period. However, tidal heating over a much longer period of time has had an even greater

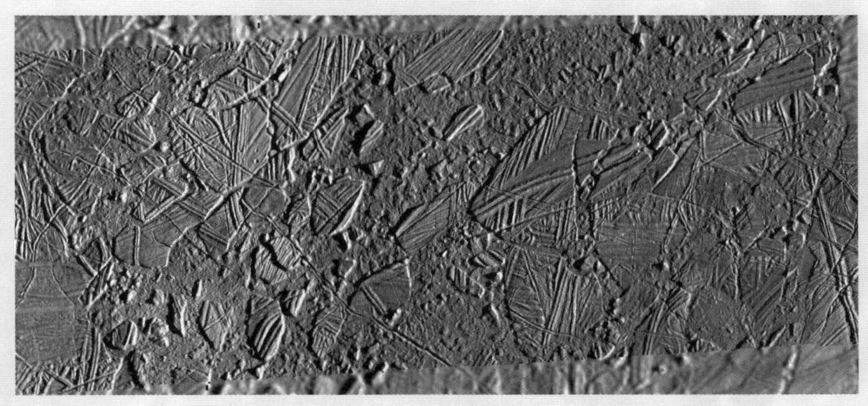

effect on Io, Europa, and to some extent Ganymede. We often think of tides as a local phenomenon of the raise and fall of the sea level against the shore. But from an astronomical point of view, tides are the distortion of bodies' shapes by the gravitational effect of other masses. Our Moon rotates synchronously, that is at the same rate as its orbital motion, so it always presents the same face toward earth. But if a moon rotates non-synchronously, its body is continually worked by its changing tidal figure. Similarly, if its orbit is eccentric rather than circular, it is continually distorted. Such deformation results in heating and stress. But for most satellites the effect is short-lived, because the tides tend to turn themselves off by

circularizing the orbit and changing the rotation to be synchronous. With a circular orbit and synchronous rotation, there is no continual tidal distortion.

Io, Europa, and Ganymede have an orbital resonance, the "Laplace relation", which maintains the eccentricities. The resonance results from the orbital periods having a ratio of 1:2:4. Their mutual gravitational effects are enhanced, with the effects of keeping the orbits eccentric and maintaining the resonance. In that way, tidal heating is maintained as well.

Tidal heating is greatest in Io because it is closest to Jupiter. It is less but still significant at Europa, but only slight at the distance of Ganymede. This trend explains the basic differences among the satellites: the great volcanic activity and removal of any water from Io; the ocean and active surface ice on Europa; and the relatively little activity evident on Ganymede. Of course, Callisto, not being involved with the resonance, shows no sign of tidal heating.

Even with the Laplace resonance, at Ganymede's distance from Jupiter the resonance is not strong enough to drive significant tidal heating. Thus the differences in internal differentiation and past geological activity between these two satellites might be due only to different formation circumstances, as discussed above.

Another possibility for extra heating of Ganymede is that the resonance may have been stronger at some time in the past. One effect of tides is to change the orbital periods of satellites. Some theories have the resonance being stronger at some time in the past, so the orbits become eccentric and tidal heating was enhanced. That scenario might explain how a burst of geological activity on Ganymede could have occurred in the first couple of billion years of the age of the satellite.

Surface appearance and geology

Europa

The continual resurfacing of the ice shell of Europa has produced two main types of terrain: tectonic (cracking and shifting of the crust) and "chaotic" (the result of heat). On a global scale (Figure 1), these features are recognizable as lines and splotches marked by small amounts of dark material, probably delivered from the ocean through cracks or melt zones. Few craters are visible because the surface is continually renewed.

■ ***Above left:*** *(Figure 5) The common denominator of ridges on Europa is the double ridge. In this densely ridged area (about 12km across at 21 m/pixel), double ridges of various sizes have formed on top of one another at various orientations. Ridge formation is a major surface renewal process on Europa.*

■ ***Above:*** *(Figure 6) A mosaic of high-resolution images (54m/pixel) of chaotic terrain within the Conamara Chaos region of Europa. Within a lumpy bumpy matrix, rafts of displaced crust display fragments of the previous tectonic terrain. Subsequent to the formation of the chaotic terrain, probably by melt-through from below, the fluid matrix refroze and new double ridges have formed across the area and begun the process of tectonic resurfacing.*

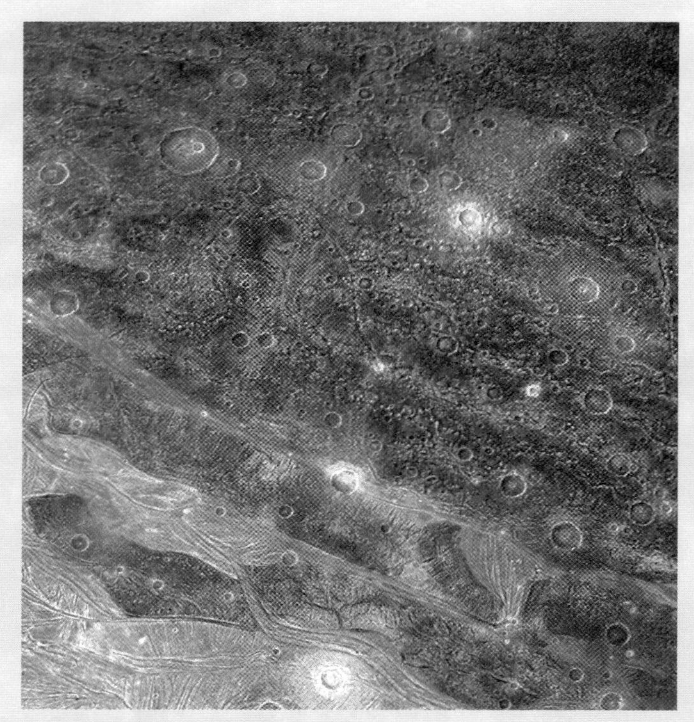

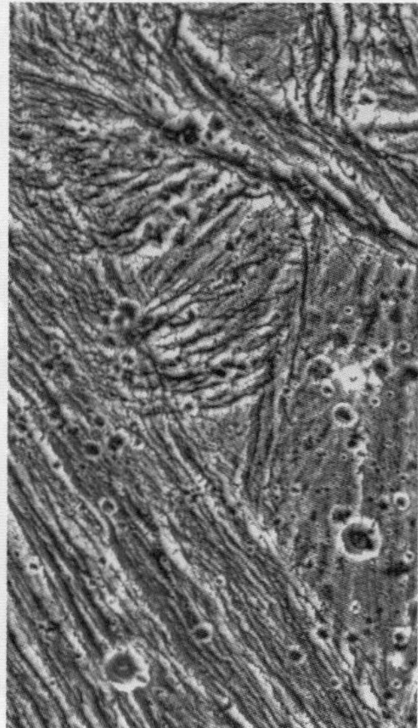

The tectonic terrain is dominated by double ridges, whose central grooves mark cracks in the crust (Figure 5). Global-scale lines often consist of multiple sets of double ridges, and they fit crack patterns expected from tidal tension. Double ridges may form as cracks initiated by tidal stress are continually worked by periodic tides that squeeze up crushed ice and slush. Many crack features have the cycloid shapes expected from tidal stress (See Figure 2), often extending ~1,000km across the surface, with each arc typically ~100km long.

Top: *(Figure 7) The dark regions on Ganymede contain sets of furrows, which are often parallel to the borders with the bright grooved terrain as seem here. These dark-region furrows may have formed by surface stretching, an incipient version of the kind of tectonic processing that produced the bright grooved terrain. Furrows in the dark regions also often display concentric patterns suggesting that they formed as rings around impact basins, similar to those of Valhalla on Callisto (Figure 4). In the foreground (lower left) is typical bright grooved terrain. Craters of various forms are displayed in this area. The area shown is about 450km wide in a Voyager image (JPL/NASA image).*

Above: *(Figure 8) High resolution image of bright, grooved terrain on Ganymede showing details of the extensional tectonics. This image, taken by the Galileo spacecraft, shows an area 35km wide at resolution 74m/pixel (JPL/NASA image).*

Large plates of the ice shell, often hundreds of kilometers across, have undergone major displacement. Often it has widened cracks, which allowed infilling from below. Plates have also often sheared past one another along cracks, a process that may be driven by daily tides. Because the observed cracks fit tidal-stress theory, the tectonic patterns record past rotation. They provide evidence for non-synchronous rotation and even for polar wander in which the icy shell slips around relative to the spin orientation of the satellite.

In the chaotic terrain the surface has been disrupted, with rafts of older terrain displaced within a lumpy matrix (Figure 6). These areas likely represent sites of thawing, followed shortly by refreezing. The ocean may melt through from below with only modest concentrations of tidal heat. While creation of chaotic terrain destroys older surfaces, the chaotic terrain in turn can be destroyed by subsequent tectonics. The history of Europa has been an on-going interplay of resurfacing, by tectonics and by chaos formation, with each destroying what was there before, and with each seemingly involving breakthrough of the ocean to the surface.

Ganymede

The darker portion (1/3) of Ganymede's icy surface is enriched in impurities, especially clay minerals, which avoided the internal differentiation and may have been accreted late in the formation of the satellite. This older terrain is covered with craters and other impact features called palimpsests and concentric furrows. Palimpsests are round, slightly brightened patches where the shape of an old crater seems to have flattened down. The concentric furrows (each typically about 10km wide and spaced 50km apart) form huge circular systems formed by giant early impacts. Other groups of furrows in the dark terrain tend to be lined up parallel to the borders with the brighter areas (Figure 7). They may be products of surface stretching (analogous to the basin and range country in the western US), the same extensional process that has created the brighter terrain.

The brighter terrain (about 2/3 of the surface, see Figures 3 and 8) consists of bands of roughly parallel grooves that have cut across the older darker surface. The grooves are probably more mature versions of the extensional furrows on the dark terrain. Extension of the icy

surface affected Ganymede very differently than Europa. On Europa, cracks have opened tens of kilometres wide, allowing bands of new material to fill in the opening, all due to the mobility provided by the near-surface ocean. There, the opposite sides of the bands match, like Africa and South America on opposite sides of the Atlantic Ocean. On Ganymede, the old terrain on opposite sides of the bright grooved terrain cannot be matched. The bright material in between is generally not new, but rather old surface that has been stretched forming the grooves and furrows. This distortion exposed some of the purer ice from just below the surface, giving the brighter appearance. In addition, some liquid water may have oozed up to the surface.

The icy surface of Ganymede has stretched so much that the whole body must have expanded since the original formation of the old, dark terrain. That global expansion may have occurred during the internal differentiation, as the lower-density ice rose up from the deep interior. As it did, the release of pressure must have allowed the ice to expand. Thus, the surface modification must have occurred at about the same time as the first major internal heating. Thus age information from the crater record can be connected with the time of heating, which is useful for developing theories of the formation and orbital evolution of the icy satellites.

Other subtle forms of change have affected Ganymede's dark terrain. Material has moved downhill yielding landslides, exposure of brighter material on slopes, and accumulation of darker material at low points. Sublimation (the evaporation of solid ice) has concentrated darker material on sun-facing slopes while frost deposits formed preferentially on the opposite slopes. Diffuse brightening near Ganymede's poles may also be explained by frost deposits.

Callisto

Callisto's surface is similar to the old, dark portion of Ganymede, consistent with the failure of the interior of Callisto to have been heated enough to differentiate fully. The surface is heavily cratered by impacts, with younger craters usually appearing bright where they have exposed purer ice below the relatively concentrated silicates at the surface. As on Ganymede, palimpsests record the locations of ancient craters, and larger impacts have left multi-ring structures of concentric cliffs and troughs, including the 3,800km wide Valhalla system (Figures 4 & 9).

The terrain has a strange softened topography with knobby protrusions, which are usually modified parts of ancient crater rims. This character has been formed by considerable down-slope motion, which has concentrated dark material in a fairly smooth blanket over the lower elevations. Sublimation has also contributed to the knobby shapes. Sublimation and frost deposits are controlled by local orientation toward the sun and by absorbtion of heating by the dark impurities at the bases of scarps and

Above: *(Figure 9) A high resolution Galileo-spacecraft view (46 m/pixel, and 33km across) of one of the concentric scarps of the Valhalla multi-ring impact structure (Figure 4). The image also shows the softened lower terrain between knobby remnants of sublimation-eroded hills and crater rims. Dark material has moved downslope leaving the relatively bright icy knobs (JPL/NASA image).*

knobs. These processes steepen slopes, creating the knobby topography and smoothing the dark terrain at the lower elevations.

Further exploration and possibilities of extraterrestrial life

Scientists generally apply lessons learned from some planets to new information about others. This approach has been reasonably successfully when dealing with rocky planets, which we can compare with a body we know well: the Earth. Icy satellites pose greater challenges. Sometimes comparison with the Earth is useful. Examples are rock structures that may be similar to cold solid ice; or icy features in the Arctic and Antarctic that can be compared with those on icy satellites; or glaciers that reveal the unique viscous qualities of ice. However, the conditions on icy satellites are so alien that such comparisons require great caution. We do not know how ice behaves in the interiors of planetary bodies, or even what the conditions are there. Similarly, the surfaces of the icy moons are colder than anyplace on Earth, and the properties of the ice and the exact compositions are uncertain. We need to be careful when we use comparisons with Earth.

At the same time, there are exciting reasons to learn more about the icy satellites. All three icy Galilean satellites probably have liquid water layers, and one, Europa, almost certainly has an ocean just below the surface. Naturally, liquid water raises the issue of extraterrestrial life. Where the liquid water is under a thick layer of ice, life would face an inhospitable setting. Any ecosystem would be isolated from both oxygen and from sunlight, and would certainly not be life as we know it.

On Europa, however, the surface geology suggests connections between the ocean and the surface, with regular flow of liquid through cracks and melt zones, and the surface may be a rich source of the necessities of life. Oxygen is separated from the H_2O ice at the surface by energetic charged particles from Jupiter's magnetosphere. Organic (carbon-rich) substances rain onto the surface from comets. Sunlight adequate for photosynthesis reaches not only the surface, but penetrates a few meters below. Organisms would need to stay a few centimeters deep to be safe from the radiation, but plenty of sunlight would still reach them and the warm sea water periodically reaching the surface could conceivably support a rich ecology, both in the crust and in the ocean. A biosphere on Europa, if any, probably extends from deep in the ocean up to within a few centimeters of the surface. On Europa, life, or its remains, may well be accessible close to the surface.

Unfortunately, most planning for exploration by NASA in the US has assumed that the search would be much more difficult, that life can only be found in the ocean. Thus exploration strategies have focused on eventually having a robot drill or melt its way down deep through thick ice. In the late 1990s, NASA anticipated a series of three missions, roughly one per decade. The first would orbit Europa, the second would land on the surface, and the third would drill down to the ocean. The first mission, the Europa Orbiter underwent two years of planning before it was cancelled due to high costs and the technical challenges of operating electronics amidst the energetic charged particles near Europa.

Hopes for a mission to the icy Jovian satellites were reborn in 2003 when NASA linked it to Project Prometheus, a program of development of nuclear electric propulsion, motivated by national defense interests as much as for planetary exploration. The centerpiece of Prometheus would be the JIMO mission, the Jupiter Icy Moons Orbiter. The icy satellites provided a scientific justification for Prometheus, while Prometheus would provide resources for exploration. Many scientists regarded the arrangement as a Faustian bargain, especially considering what ultimately happened to the original Prometheus. Within a couple of years, this project was also cancelled due to budget constraints.

NASA continues to identify Europa is its highest priority objective for exploration in the outer solar system. That status may not have much meaning, given that US resources are becoming extremely limited and the president has declared the Moon and Mars to be of highest priority. However, at the same time, the European Space Agency and other national space agencies have begun to consider exploration strategies of their own for the Jovian satellites.

A key to success may be that life, or its remains, could be readily accessible close to the surface, at least on Europa. Thus mission planners should aim to land in a promising location, rather than assume that deep drilling is needed to reach their objective. They also need to take seriously the possibility that spacecraft from Earth could contaminate with terrestrial organisms any moon, like Europa, with a habitable zone near its surface.

In preparation for any such venture, comparative studies of all icy satellites will be essential. Certainly, the possibility of finding life makes further investigation of the icy satellites especially exciting. But even if extraterrestrial life were out of the question, the remarkable properties and processes on these complex worlds make them worthy of continuing exploration.

Acknowledgment: The author thanks NASA's Outer Planet Research program for support of this work.

__Richard Greenberg__ received his PhD from the Massachusetts Institute of Technology and is Professor of Planetary Science at the University of Arizona. He was a member of the imaging team for the Galileo mission to Jupiter and he wrote **Europa – the ocean moon** *(Springer Praxis books). His current research includes the origin and dynamic evolution of planets and satellites, the effects of tides on the geophysics of natural satellites and the geology of Europa.*

Further reading:

Anderson, J.D., et al. (2005) Amalthea's density is less than that of water, Science 308, 1291-1293.

Greenberg, R. (1989) Time-varying orbits and tidal heating of the Galilean satellites. In *Time-Variable Phenomena in the Jovian System* (Belton, West, and Rahe, eds.), NASA Special Publication 494.

Greenberg, R. (2005) *Europa, the Ocean Moon*, Springer-Praxis, NY.

Hoppa, G.V., et al. (1999) Formation of cycloidal features on Europa. Science 285, 1899-1902.

Kivelson, M.G., et al. (2000) Galileo magnetometer measurements: A stronger case for a subsurface ocean at Europa, Science 289, 1340-1343.

Bagenal, F., et al., eds. (2004) *Jupiter: The Planet, Satellites, and Magnetosphere*, Cambridge University Press. See chapters on "The origin of Jupiter", "Callisto", "Geology of Ganymede", "Interior composition, structure, and dynamics of the galilean satellites".

Squyres, S.W., et al. (1983), Liquid water and active resurfacing on Europa, Nature, *301*, 225-226.

Below: *Two views of the trailing hemisphere of Jupiter's ice-covered satellite, Europa. The left image shows the approximate natural colour appearance of Europa. The image on the right is a false-colour composite version combining violet, green and infrared images to enhance colour differences in the predominantly water-ice crust of Europa. Image courtesy NASA/Jet Propulsion Laboratory, with thanks to Deutsche Forschungsanstalt für Luft und Raumfahrt e.V., Berlin, Germany.*

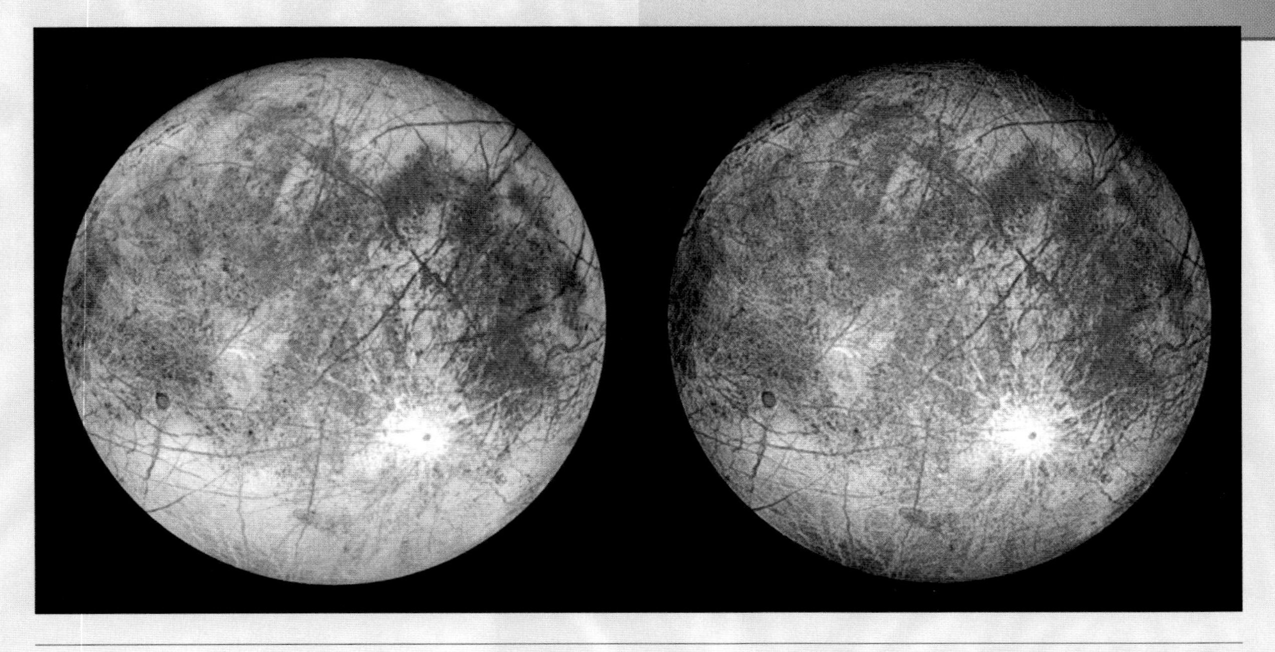

4.4

Exploring the RINGED PLANET

It is now two years since the Cassini spacecraft arrived in orbit around the ringed planet Saturn. Here, *David Harland* reviews the results so far...

Cassini at SATURN

THE CASSINI-HUYGENS mission to Saturn was developed by the American space agency NASA, the European Space Agency (ESA) and the Italian Space Agency (ASI). At 5.7 tonnes, the two-part spacecraft is the largest, most sophisticated deep space probe to-date. It was launched on 15 October 1997 and flew an interplanetary cruise that used gravitational 'slingshots' with Venus, Earth and Jupiter in order to reach Saturn. The highlights of the early part of its mission in the Saturnian system were close observations of the satellites Phoebe, Titan, Iapetus, Hyperion and Enceladus but the craft, which is performing flawlessly, has also undertaken long-term investigations of the ring system and the planet's atmosphere and magnetosphere.

Phoebe

A 2,071km flyby of the outer moon Phoebe on 11 June 2004 assisted in braking the spacecraft, thereby saving propellant for use later in extending the tour of the Saturnian system. Phoebe, which had not been well situated for study by the Voyagers, was able to be imaged with a resolution of several tens of metres. The 200km body is intensely cratered, with a spectrum similar to that of a C-type (carbonaceous) asteroid and so is an unprocessed relic of the solar nebula. Its retrograde orbit in the plane of the ecliptic implies that it is a captured object from beyond the orbit of Neptune.

Above: *Saturn's moon Phoebe is revealed in this mosaic of two images taken during Cassini's flyby on 11 June 2004. Phoebe may be an ice-rich body coated with a thin layer of dark material. Image courtesy NASA/JPL/Space Science Institute.*

On 27 June, Cassini encountered the 'bow shock', where the solar wind was compressed against the sunward boundary of the planet's magnetosphere. The fact that this feature was at a distance of 49 planetary radii was a surprise, as when the Voyagers had flown by it had been at half this distance from the planet. The next day, the spacecraft entered the calmer environment of the magnetosphere. Whereas the intense magnetic field of Jupiter is able to trap a vast number of energetic ions ('plasma') within its magnetosphere, Saturn's magnetic field is much weaker and its magnetosphere is populated mainly by electrons and neutral gas; only its outer part contains plasma and this is derived from the solar wind.

Arrival

At 02:36 GMT on 1 July at mission control, the Jet Propulsion Laboratory (JPL) in California, it was late the previous evening, shortly after Cassini crossed the ring plane by passing through the 'gap' between the F- and G-rings, it initiated the 95min insertion burn and this braked the spacecraft into a highly elliptical 'capture orbit' whose parameters were so close to those scheduled

dark material, as there was on Phoebe. Nevertheless, at visible wavelengths the dark material on Iapetus resembles a D-type (reddish) asteroid – as indeed does Hyperion – whereas Phoebe is more like the C-type asteroid.

Releasing Huygens

On its way back in, on 26 October, Cassini had its first close encounter (designated Ta) with Titan. Despite the point of closest approach being 1,174km above the surface of the moon, the spacecraft skimmed the fringe of its atmosphere and the inlet of the mass spectrometer collected a sample of the material for chemical analysis. The second flyby (Tb) occurred on 13 December and was at the greater altitude of 2,358km. The results of these encounters confirmed the validity of the atmospheric model used in planning the Huygens mission. This probe was released on 25 December, on a collision course with Titan. The spring-ejector spun the 319kg probe at seven revolutions per minute for stability and then pushed it away from Cassini at 30cm/sec.

that the optional trim burn was deleted. Images taken during the insertion sequence showed the elusive moonlet Pan which orbits within the Encke Gap, near the outer edge of the A-ring. As the spacecraft climbed away from the planet, it inspected Titan from a range of 340,000km, glimpsing the first detail on the surface of this cloud-enshrouded moon and a long-range view of Iapetus revealed for the first time some surface detail on the mysterious dark Cassini Regio of the moon's leading hemisphere, establishing there to be a number of large impact basins. The spacecraft fired its engine again on 23 August to raise its periapsis clear of the ring system. The apoapsis of the capture orbit occurred on 27 August, at a distance of 151 planetary radii.

Iapetus from afar

On 18 October 2004, Cassini passed within 1.1m km of Iapetus and revealed there to be a line of steep-sided mountains running across Cassini Regio. Four days later, when the phase of the moon was a narrow sunlit crescent, a long-exposure image captured the dark side illuminated by 'Saturnshine'. Although the resolution of the visual and infrared mapping spectrometer was such that its image of Iapetus spanned just four pixels, the fact that one pixel was on the bright terrain and another was on Cassini Regio offered a basic comparison of the compositions of the two terrains. This showed that there was carbon dioxide in the

The only active component of the probe was the timer that would awaken it as it approached its objective 20.3 days later.

Iapetus up close

On 27 December 2004, Cassini fired its engine to produce a 60,000km flyby of Titan (Tc), on a trajectory with a clear line of sight to receive the transmission from the probe as this penetrated the moon's atmosphere. The two vehicles passed apoapsis at 60 planetary radii on 31 December. On that same day Cassini observed

Above: *An artist's impression of Cassini during the Saturn Orbit Insertion(SOI) manoeuvre, just after the main engine has begun firing to reduce the spacecraft's velocity with respect to Saturn. Image courtesy NASA/JPL.*

Right: *An artist's impression of the Cassini spacecraft releasing the Huygens probe on 24 December 2004 on a trajectory that would take it to Titan. Image courtesy NASA/JPL.*

Iapetus from 123,400km and it was discovered that the peaks noted earlier were part of a narrow 13km-tall ridge of bright icy material that extends all around the moon, including across Cassini Regio, giving the moon the appearance of a walnut. The fact that the ridge lies on the equator suggests that it may be a re-accreted ring of ejecta thrown up by a large impact.

Despite our having finally glimpsed the topography on Cassini Regio, it is still not evident whether the dark material is of endogenic or exogenic origin. Regardless of its origin, the fact that the material lies on top of other features means it is a coating acquired after the bombardment that cratered the moon's surface. The uniform appearance of the dark material in the equatorial zone, its apparent thinning and spottiness at higher latitudes and dark wispy streaks near the distal margin of Cassini Regio, suggest that this material was ballistically deposited. One theory is that since Phoebe has a retrograde orbit (as indeed do several moonlets farther out that might be related to Phoebe), sputtered dust motes will tend to spiral into the system and be swept up by the leading edge of the next moon in, which is Iapetus. In one endogenic-origin theory, methane was erupted onto the surface, where it was

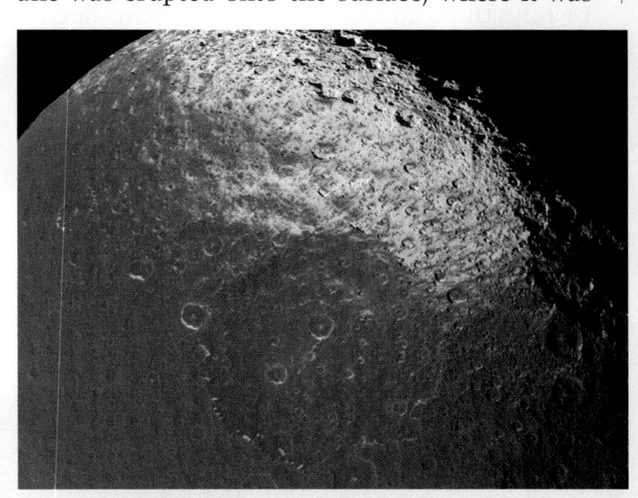

subsequently darkened by solar ultraviolet. As there is no evidence of fluid flows across the dark surface – although there are spectacular landslides – such an eruption may have taken the form of a plume. But what process drove this volcanism and why was it confined to the moon's leading hemisphere?

Huygens and Titan

On 14 January 2005, Huygens penetrated Titan's atmosphere at 21,000km/hr (some 6km/sec), its 2.7m diameter conical heat shield protecting it

from the 1,700°C entry temperature. Having shed its shield, the probe performed a nominal parachute descent lasting 2.5hr and then, after touching down with a 15-*g* jolt, to the amazement of all concerned, it continued to function on the surface for an hour, its transmission being curtailed only when Cassini was obliged to turn its 4m-diameter antenna away in order to perform other tasks. Although a fault in the communications system had been overcome by redesigning Cassini's flyby to limit the Doppler effect, an error in programming meant that one of the two signal channels from the probe was lost, which cost roughly half of the imagery and, unfortunately, the facility for the Doppler wind experiment (although some of this science was recoverable from monitoring of the probe's carrier signal by terrestrial antennas).

By any measure, the probe's entry, descent and landing greatly exceeded expectations. Although the probe had been designed to float in a cryogenic liquid hydrocarbon brew of methane, propane and butane, it landed on a plain littered with small icy pebbles that was strongly suggestive of a dry lake bed. On Titan, methane can exist as a solid, a liquid and a gas and can play the same role as water in a hydrological cycle, with drainage of rain fall eroding the surface. The images taken by the probe as it descended showed networks of dark channels draining down off the bright terrain onto the dark area. Although the landing site was dry, the context implies that the occasional 'deluge' gives rise to a flash flood. The probe's monitor-

Left: *This Cassini of image Saturn's intriguing moon Iapetus shows the northern part of the dark Cassini Regio and the transition zone to a brighter surface at high northern latitudes. Image courtesy NASA/JPL/Space Science Institute.*

Above: *An artist's impression depicting the parachute descent of the Huygens lander through Titan's atmosphere to its eventual landing on the icy surface. Image courtesy C. Carreau and the European Space Agency.*

ing of the site suggests that methane was present just beneath the surface, as if the lake had recently soaked into the porous icy surface material rather than evaporated into the atmosphere. Despite the fascinating results – which have provided the 'ground truth' necessary to interpret the remote-sensing by the mother ship – it will be a very long time before another probe lands on Titan and thus, for the foreseeable future, Huygens' view of its surface is sure to become one of those iconic space images.

Enceladus

A major objective was to investigate the enigmatic 500km-diameter moon, Enceladus, which, by reflecting some 90% of incident light, has the highest albedo of any object in the solar system. The Voyagers revealed its surface to be partly cratered and partly smooth, as if resurfaced. Might it be undergoing a form of cryovolcanism? After its 1,600km flyby of Titan on 15 February, Cassini passed Enceladus at a range of about 1,200km on 17 February and on its next orbit made a 500km flyby on 9 March.

During the first encounter, the magnetometer noted a striking deflection in the planetary magnetic field and this was more striking on the second, closer flyby. This meant there was a plasma in the immediate vicinity of the moon and, in fact, there were indications of ionized molecules of water vapour. Since Enceladus is too small for its weak gravitational field to retain an atmosphere, it was evident that this must emerge from the surface and leak to space, where it forms the diffuse E-ring, within which the moon's orbit lies. Belatedly, it was realized that a large cloud of oxygen observed in the E-ring soon after Cassini's arrival in the Saturnian system and which dissipated over the next several months, must have originated from Enceladus.

Cassini made another pass by Enceladus on 14 July, this time at a range of 172km, during which a number of parallel fractures 130km in length and spaced 40km apart, which were immediately dubbed 'tiger stripes', were discovered in the south polar region. Furthermore, the moon's south pole is anomalously warm – whereas the temperature at the equatorial zone is 80°K and the south pole averages 85°K, there are small areas near these fractures at 110°K. In addition, the cloud of water vapour noted previously was now found to be concentrated in the south polar region.

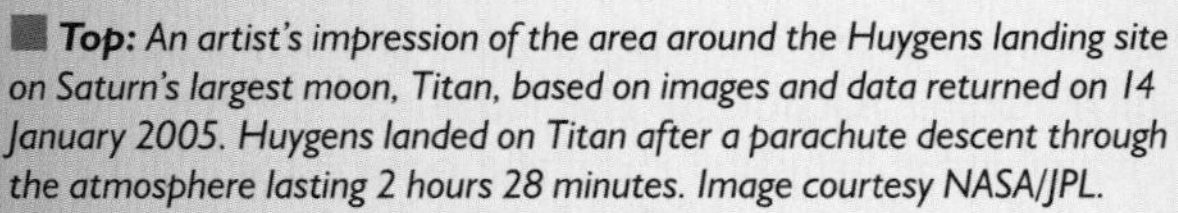

Top: *An artist's impression of the area around the Huygens landing site on Saturn's largest moon, Titan, based on images and data returned on 14 January 2005. Huygens landed on Titan after a parachute descent through the atmosphere lasting 2 hours 28 minutes. Image courtesy NASA/JPL.*

Above: *As it passed Saturn's moon Enceladus on 14 July 2005, Cassini acquired high resolution views of this puzzling ice world. Enceladus exhibits a bizarre mixture of softened craters and complex, fractured terrain. Many of the long fractures exhibit a pronounced difference in colour (shown here in blue) from surrounding areas. Image courtesy NASA/JPL/Space Science Institute.*

Right: *Cassini images of Saturn's moon Enceladus, backlit by the Sun, show fountain-like jets emanating from the south polar region. It is thought that the jets are geysers erupting from pressurized subsurface reservoirs of liquid water. Image courtesy NASA/JPL/Space Science Institute.*

Was this anomalous zone the source of the vapour? This was confirmed when a number of 'jets' were photographed spewing plumes of vapour hundreds of kilometres into space. The absence of ammonia and the sheer amount of vapour, suggests pure-water volcanism. The question now is: what is the heat source that is driving this activity?

Hyperion

On 26 September 2005, Cassini flew within 514km of Hyperion, the highly irregular moon whose rotation is chaotic. Its low density implies it is mostly made of water ice. At 164 by 130 by 107km, this object is near the size limit at which internal pressure due to its gravitational field will begin to crush ice, closing pore spaces and creating a spheroidal shape, but not sufficiently massive for this to have taken effect. The flyby confirmed it to have a spongey structure, suggesting it is little more than a pile of icy rubble. The dark material on the surface is probably a minor constituent and possibly of exogenic origin.

Rings and things

One objective of the mission is to study the ring system to determine whether it is permanent or temporary. The ring system had changed little since the Voyagers, but Cassini's instruments could observe at a wider range of wavelengths and with greater resolution. The gravitational perturbations of the moonlets within and just beyond the ring system cause density waves within the rings, 'scallop' their edges and warp them out of plane. A stream of material was seen being drawn from the F-ring towards its inner 'shepherd', Prometheus. When stars were occulted by the rings, measurements of how the light 'flickered' established that the ringlets have very sharply defined edges. The gravitational effects of the shepherding moonlets enabled their masses to be inferred and the densities of 0.5gm/cm^3 suggest they are loosely consolidated bodies rather than solid ice. Cassini found that whilst the rings are water-ice, in places this is 'contaminated' by dark material with a similar composition to Phoebe. One hypothesis says that the more contaminated a ring, the older it is. This strengthened the case for the system being the relic of a disrupted moon and thus only an ephemeral feature that will last only a few hundred million years.

Top: *This six-image mosaic of Saturn's heavily cratered moon Hyperion, acquired by Cassini on 26 September 2005, reveals a low-density body, blasted by impacts over eons, giving it a spongey appearance. Image courtesy NASA/JPL/Space Science Institute.*

Middle left: *Dark, diagonal drapes in the inner strands of icy particles in Saturn's F-ring caused by the gravitational influence of the shepherd moon Prometheus, which orbits inside the ring. Another shepherd moon, Pandora, orbits outside of the ring. Image courtesy NASA/JPL/Space Science Institute.*

Bottom left: *Here, colour is used to indicate ring particle sizes in different regions based on the measured effects for three radio signals. Purple indicates a lack of particles less than 5cm across. Green and blue indicate regions where there are particles smaller than 5cm and 1cm, respectively. All ring regions appear to be populated by a broad range in particle sizes that extends to boulder-sized fragments, many metres across. Image courtesy NASA/JPL.*

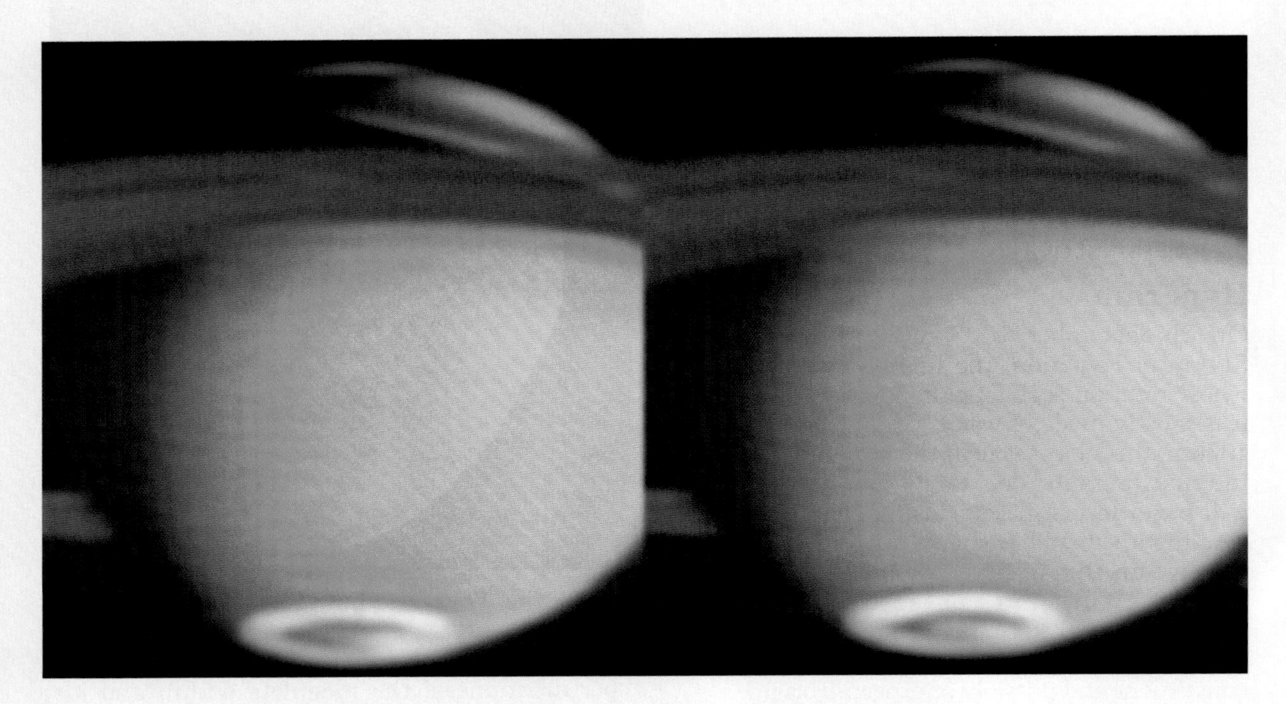

As regards Saturn itself, a major increase in auroral activity in the polar regions on 25 July 2004 was marked when a burst of solar wind penetrated the magnetosphere. Although the atmosphere is normally fairly quiescent, in the months after Cassini's arrival several major storm systems occurred in the upper atmosphere in a belt at latitude 36° south, with wind speeds of 1,750km per hour. Also, resolving a question left over from the Voyagers, it was confirmed that lightning in storms produces powerful broadband electromagnetic bursts.

Another priority is to sample Saturn's 'magnetotail', where the magnetosphere is stretched out by the solar wind. By making the flybys of Titan in different ways, it is possible to control the spacecraft's orbit. Late in the four-year primary mission, a highly eccentric orbit with its apoapsis down-Sun will be set up. Also, because the planetary atmospheres researchers require whole-disk studies of Saturn under high illumination, a highly eccentric orbit will be set up with its apoapsis up-Sun of the planet. In effect, it will be necessary to rotate the apoapsis through 180°. At one point in this sequence, the inclination of the orbit and its apoapsis will permit a close flyby of Iapetus. By the end of the primary mission, the spacecraft's orbit will be inclined at 84° and this will permit direct observations of the polar regions of the planet and its magnetosphere, provide strikingly 'open' views of the ring system and opportunities to use stellar occultations to profile the distribution of material within the rings.

Although the primary mission will be formally concluded on 1 July 2008, this is merely a funding milestone, because if, as seems likely, Cassini is still operational, its tour will almost certainly be extended for follow-up studies.

Further reading:

David M Harland: *Mission to Saturn - Cassini and the Huygens probe.* Springer-Praxis, 2002.

Dr David M Harland *gained his BSc in astronomy in 1977 and a doctorate in computational science. He has subsequently taught computer science, worked in industry and managed academic research. In 1995 he retired and has since published many well known books on space themes.*

Above: *Cassini spacecraft images of oval-shaped auroral emissions around Saturn's south pole. In the side-by-side, false-colour images, blue represents aurora emissions from hydrogen gas excited by electron bombardment, while red-orange represents reflected sunlight. Image courtesy NASA/JPL/University of Colorado.*

Table: Cassini's primary mission, as planned at the start of the tour.

Date (GMT)	Event		Range (km)	
11 Jun 2004	**Phoebe**	P1	**2,071**	*Inbound*
1 Jul 2004	**SOI**	-	**-**	-
2 Jul 2004	**Titan**	-	**340,000**	*Outbound*
23 Aug 2004	**PRM**	-	**-**	-
27 Aug 2004	**Apoapsis**	-	**-**	$R=151R_S$
26 Oct 2004	**Titan**	Ta	**1,174**	*In*
13 Dec 2004	**Titan**	Tb	**2,358**	*In*
15 Dec 2004	**Dione**	-	**72,000**	*In*
15 Dec 2004	**Mimas**	-	**107,000**	*In*
25 Dec 2004	**Huygens release**	-	**-**	*Out*
27 Dec 2004	**ODM**	-	**-**	-
31 Dec 2004	**Apoapsis**	-	**-**	$R=60R_S$
31 Dec 2004	**Iapetus**	-	**123,400**	*In*
14 Jan 2005	**Titan (probe relay)**	Tc	**60,000**	*In*
16 Jan 2005	**Mimas**	-	**108,000**	*In*
15 Feb 2005	**Titan**	T3	**1,600**	*In*
17 Feb 2005	**Enceladus**	E1	**1,264**	*Out*
9 Mar 2005	**Enceladus**	E2	**500**	*In*
9 Mar 2005	**Tethys**	-	**83,000**	*In*
29 Mar 2005	**Tethys**	-	**108,000**	*In*
29 Mar 2005	**Enceladus**	-	**55,580**	*In*
31 Mar 2005	**Titan**	T4	**2,400**	*Out*
15 Apr 2005	**Mimas**	-	**82,500**	*Out*
16 Apr 2005	**Titan**	T5	**1,026**	*Out*
2 May 2005	**Tethys**	-	**52,000**	*In*
21 May 2005	**Enceladus**	-	**102,000**	*Out*
14 Jul 2005	**Enceladus**	E3	**172**	*In*
2 Aug 2005	**Mimas**	-	**63,000**	*In*
22 Aug 2005	**Titan**	T6	**3,700**	*Out*
7 Sep 2005	**Titan**	T7	**1,075**	*Out*
23 Sep 2005	**Mimas**	-	**70,000**	*In*
24 Sep 2005	**Tethys**	T1	**1,500**	*Out*
26 Sep 2005	**Hyperion**	H1	**514**	*Out*
11 Oct 2005	**Dione**	D1	**500**	*In*
12 Oct 2005	**Enceladus**	-	**49,000**	*Out*
28 Oct 2005	**Titan**	T8	**1,353**	*In*
26 Nov 2005	**Rhea**	R1	**500**	*In*
27 Nov 2005	**Enceladus**	-	**108,000**	*In*
24 Dec 2005	**Enceladus**	-	**94,000**	*In*

Table: Cassini's primary mission, as planned at the start of the tour. (cont'd)

Date (GMT)	Event		Range (km)	
26 Dec 2005	**Titan**	T9	**10,400**	*Out*
15 Jan 2006	**Titan**	T10	**2,043**	*In*
27 Feb 2006	**Titan**	T11	**1,813**	*Out*
19 Mar 2006	**Titan**	T12	**1,951**	*In*
30 Apr 2006	**Titan**	T13	**1,855**	*Out*
20 May 2006	**Titan**	T14	**1,879**	*In*
2 Jul 2006	**Titan**	T15	**1,906**	*Out*
22 Jul 2006	**Titan**	T16	**950**	*In*
7 Sep 2006	**Titan**	T17	**950**	*In*
23 Sep 2006	**Titan**	T18	**950**	*In*
9 Oct 2006	**Titan**	T19	**950**	*In*
25 Oct 2006	**Titan**	T20	**950**	*In*
12 Dec 2006	**Titan**	T21	**950**	*In*
28 Dec 2006	**Titan**	T22	**1,500**	*In*
13 Jan 2007	**Titan**	T23	**950**	*In*
29 Jan 2007	**Titan**	T24	**2,726**	*In*
22 Feb 2007	**Titan**	T25	**950**	*Out*
10 Mar 2007	**Titan**	T26	**950**	*Out*
26 Mar 2007	**Titan**	T27	**950**	*Out*
10 Apr 2007	**Titan**	T28	**950**	*Out*
26 Apr 2007	**Titan**	T29	**950**	*Out*
12 May 2007	**Titan**	T30	**950**	*Out*
28 May 2007	**Titan**	T31	**2,426**	*Out*
13 Jun 2007	**Titan**	T32	**950**	*Out*
29 Jun 2007	**Titan**	T33	**1,944**	*Out*
19 Jul 2007	**Titan**	T34	**1,300**	*In*
31 Aug 2007	**Titan**	T35	**3,212**	*Out*
10 Sep 2007	**Iapetus**	I1	**1,229**	*Out*
2 Oct 2007	**Titan**	T36	**950**	*Out*
19 Nov 2007	**Titan**	T37	**950**	*Out*
5 Dec 2007	**Titan**	T38	**1,300**	*Out*
20 Dec 2007	**Titan**	T39	**950**	*Out*
5 Jan 2008	**Titan**	T40	**950**	*Out*
22 Feb 2008	**Titan**	T41	**950**	*Out*
12 Mar 2008	**Enceladus**	E4	**97**	*In*
25 Mar 2008	**Titan**	T42	**950**	*Out*
12 May 2008	**Titan**	T43	**950**	*Out*
28 May 2008	**Titan**	T44	**1,350**	*Out*
31 Jul 2008	**Titan**	T45	**3,980**	*Out*

While cruising around Saturn in early October 2004, the Cassini spacecraft captured a series of images that have been composed into a detailed, global natural colour view of Saturn and its rings. The images were acquired on 6 October 2004, while Cassini was approximately 6.3m km from the planet. Image courtesy NASA/JPL/Space Science Institute.

4.5 Secrets of a CLOUDY Moon . . .

The landing of the European space probe Huygens on Saturn's largest moon, Titan, was one of the most remarkable achievements of robotic space exploration. Here, *Rosaly Lopes* previews this historic landing by describing how concerted observations by the Cassini Orbiter are revealing Titan's secrets.

TITAN

Cassini-Huygens reveals A NEW WORLD

THE YEAR 2005 will be remembered in the history of space exploration for the first landing of a probe on a surface in the outer solar system - on 14 January, the Huygens probe landed on the surface of the mysterious moon Titan. The landing of Huygens on Titan was a remarkable achievement. For comparison, let's consider landing on Mars, whose distance from Earth is only 55m to 401m km, a planet explored by numerous spacecraft, starting with Mariner 4 in 1965. We successfully landed two Viking probes on Mars in the 1970s, followed by Pathfinder, Spirit, and Opportunity. We have studied Mars with numerous orbiters and fly-bys since Mariner 4, yet we still consider a successful landing on the Red Planet to be no small feat. Now consider Titan, about 1,300m km away from the Earth. The Huygens probe was built before Titan's surface had ever been mapped, before any spacecraft had ever orbited this planet-sized moon to make measurements considered necessary for safe descent and landing. Mars has a very tenuous atmosphere, in contrast, Titan's atmosphere extends 10 times further into space than the Earth's and the pressure at the surface is 50% higher. A body's atmosphere has a significant effect on probe entry and descent, yet Huygens had to make its way using scarce knowledge of atmospheric conditions. Surface conditions were also virtually unknown. Given the possibility of pools of liquids, Huygens could have landed (or crashed) into either liquid or solid. Great engineering and science and undoubtedly some luck, ensured that the probe reached its destination, landed softly and obtained remarkable measurements both during descent and while on the ground.

What did we know about Titan before Cassini and Huygens? Saturn's largest moon was discovered in 1655 by Dutch astronomer Christiaan Huygens. At 5,150km in diameter, Titan is larger than the planets Mercury and Pluto, and is the second largest known moon (only Ganymede is larger). Its atmosphere is the second densest of the solid bodies in the solar system (Venus has the densest), and its surface, shrouded by a thick atmosphere, is extremely cold, about 95° K (-178°C). Titan's thick atmosphere is about 95% nitrogen, with a few percent methane. Titan is of special interest to planetary scientists for several reasons, one of which is the many interesting organic compounds created by ultraviolet light from the Sun and energetic electrons from Saturn's radiation belts causing dissociation of nitrogen and methane in the upper atmosphere. The resulting radicals combine to form complex hydrocarbons such as ethane and benzene, nitriles, and Titan's orange haze, which obscures the surface. When the Voyager 1 spacecraft flew by Titan in 1981, its cameras showed an orange ball, much to the disappointment of planetary geologists (1,2). The Cassini orbiter instruments were designed to penetrate the haze and image the surface (3). Specifically, the Titan Radar Mapper (4) is part of the payload, it is able to reveal the surface in unprecedented detail. Two other instruments have contributed to our current knowledge of Titan's surface: the Imaging Science Subsystem (ISS, 5) can use filters that cut through some of the haze, and the Visible and Infrared Mapping Spectrometer (VIMS, 6) can make use of methane spectral "windows" to see down to the surface. Other orbiter instruments contribute to our knowledge of Titan's atmosphere and interaction with the solar wind and with Saturn's magnetosphere. The Cassini mission and its payload of instruments are reviewed in Russell et al. (3) and Harland (7). The knowledge of Titan before Cassini and Huygens is well summarized in the books by Coustenis and Taylor (1) and Lorenz and Mitton (2).

The Cassini-Huygens mission

Cassini-Huygens is a collaborative mission between NASA, the European Space Agency and the Italian Space Agency. It has the participation of 17 countries, making it the most international planetary mission ever flown. The orbiter carries 12 instruments and makes repeated fly-bys of Titan, as well as of other moons. Cassini was launched on 15 October 1997, and inserted into orbit around Saturn on 1 July 2004. The Huygens probe, carried by the orbiter, was released on 24 December 2004, and landed on Titan's surface on 14 January 2005 (8). The orbiter's first fly-by of Titan was on 2 July, 2004, at a distance of over 300,000km. Closer fly-bys followed, some as low as 1,150km, using different instruments, and Titan begun to yield its secrets. No doubt many more will be revealed in the next few years. The Cassini mission is scheduled to last until at least mid-2008, making a total of 44 fly-bys of Titan.

Titan's atmosphere: initial results from the Cassini orbiter

The photochemical processes in Titan's atmosphere are thought to be similar to those in the early Earth and it is likely that understanding the atmospheric processes on Titan will contribute to knowledge of the Earth prior to the evolution of life. One of the key objectives of the Cassini-Huygens mission is to understand Titan's atmospheric dynamics and chemistry.

The organic compounds that are photochemically produced in the atmosphere eventually condense and rain down to the surface. Titan has methane rain or possibly, as Cassini scientist Ralph Lorenz suggested, methane monsoons (9), a term first coined by Arthur C. Clarke in his book "Imperial Earth". One of the mysteries of Titan is the amount of methane in the atmosphere. Because methane forms the basis of many photochemical reactions on Titan, it should have been depleted over time, but this has not happened. What is the mechanism re-supplying the methane? Planetary scientist Jonathan Lunine proposed an ocean of liquid hydrocarbons (10), but this has not been observed. Current results suggest that surface liquids are not plentiful enough, so there must be some other process going on. Volcanism is a possibility, as several possible volcanic features have been seen on the surface (4, 10, 11), though there is not evidence that they are still active.

Cassini carries a mass spectrometer that can sample Titan's surface during close fly-bys and determine molecular masses of the atmospheric constituents. During the first close fly-by in October 2004, a complex array of short and long-chain hydrocarbons and nitriles was detected (12). One of the most significant findings is that Titan's atmosphere is enriched in the heavy isotope of nitrogen (^{15}N) relative to the more abundant ^{14}N (12). The heavier isotope is much more abundant on Titan, relative to the lighter one, than it is on Earth. What is causing this enrichment? Another of Cassini's instruments, the Magnetospheric Imaging Instrument (MIMI) detected a gas torus around Saturn during orbit insertion (13). The torus circles Saturn along with Titan, implying that Titan is the source. Titan's atmosphere is continuously bombarded with energetic electrons from Saturn's radiation belts. The collisions cause gas from Titan's atmosphere to be ejected; the lighter isotopes can move faster and escape more easily from Titan's gravitational pull. It is thought that vast

■ **Above:** *Fig 1: Cassini has found Titan's upper atmosphere to consist of a surprising number of layers of haze, as shown in this ultraviolet image of Titan's night side limb, colourized to look like true colour. The many fine haze layers extend several hundred kilometres above the surface. Although this is a night side view, with only a thin crescent receiving direct sunlight, the haze layers are bright from light scattered through the atmosphere. Image courtesy NASA/JPL/Space Science Institute.*

amounts of Titan's atmosphere must have been lost to result in the ^{15}N enrichment we see today. According to Hunter Waite and colleagues (12), the atmosphere was, in the past, between 1.6 and 100 times more massive than today.

Titan's atmosphere (Fig. 1) is dynamic, perhaps even stormy. ISS and VIMS images showed methane clouds near the south pole (5, Fig. 2). Cassini observed Titan while it was summer in the southern hemisphere, so the relative abundance of clouds in the south polar region may be a result of evaporation of surface methane. The clouds have been seen to move between different fly-bys, or even different images in the same fly-by. On some occasions they are not present at all. Other clouds have been detected, not only from Cassini but also from ground-based observations. Mid-latitude clouds seem to last only a few hours and are fainter than those at the south pole. There are some mysterious east-west clouds that are hundreds of kilometres long, streaky in appearance, and seem to originate from fixed positions. These have been speculated to be a result of venting of gases from the interior, which are then carried along by winds (14).

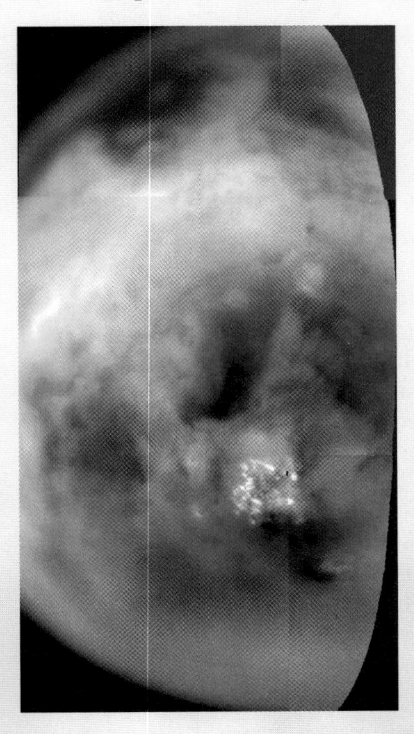

Titan's surface: the view from the orbiter

Cassini's first views of Titan's surface came on 2 July 2004. Images from ISS and VIMS showed distinct light and dark regions (5, 9), which are suspected of having different compositions. One possibility is that darker areas at visible wavelengths have more organic material, while brighter areas may have more water ice. The largest bright area, Xanadu, had been known from Earth-based infrared observations prior to Cassini. This area is thousands of kilometres wide and has often been thought of as being of higher elevation, but observations have not yet confirmed this. Radar observations obtained in four fly-bys to date (October 2004 and February, September, and October 2005) have shown the surface geology in detail over about 4% of Titan's

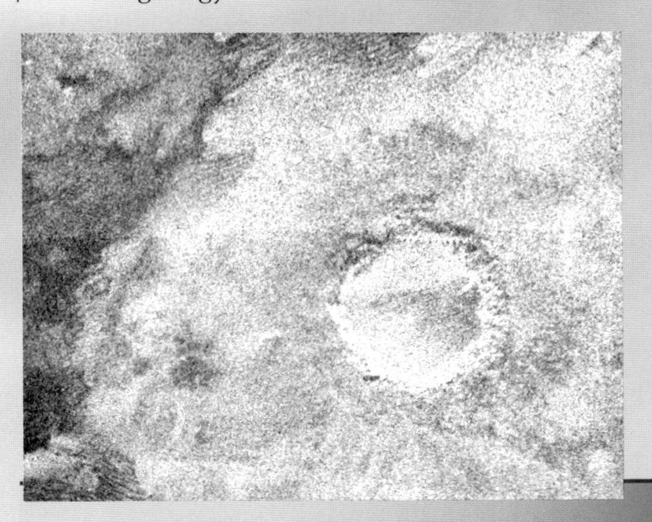

Above left: *Fig. 2: Mosaic of images of Titan's south polar region acquired as Cassini passed by at a range of 339,000km on 2 July 2004. These images were acquired through special filters designed to see through the thick haze and atmosphere. The bright spots near the bottom represent a field of clouds near the south pole. Image courtesy NASA/JPL/Space Science Institute.*

Opposite: *Fig. 3: This image taken by Cassini's visual and infrared mapping spectrometer (VIMS) is a composite of false-colour images taken at three infrared wavelengths: 2 microns (blue); 2.7 microns (red); and 5 microns (green). Surface features can be seen, as well as a methane cloud at the south pole (bottom of image). This picture was obtained as Cassini flew by Titan at altitudes ranging from 100,000km to 140,000km. The inset picture shows the area of the landing site of Cassini's Huygens probe. Image courtesy NASA/JPL/University of Arizona.*

Above right: *Fig. 4: This radar image shows one of the two impact craters detected on Titan to date. This crater, approximately 80km in diameter, on the very eastern end of the radar image strip taken by the Cassini orbiter on its third close flyby of Titan on 15 February 2005. The appearance of the crater and the extremely bright (hence rough) blanket of material surrounding it is indicative of an origin by impact, in which a hypervelocity comet or asteroid, in this case, roughly 5-10km in size, slammed into the surface of Titan. The bright surrounding blanket is debris, or ejecta, thrown out of the crater. Image courtesy NASA/JPL.*

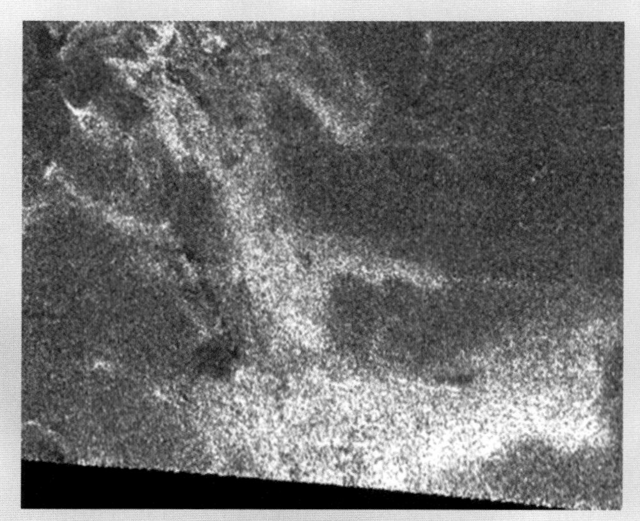

surface (4, 15). Because the Synthetic Aperture Radar mode can only be used when the spacecraft is relatively close to the surface, the area imaged in each fly-by is small, around 1%.

Titan's surface has shown itself to be remarkably varied and, apparently, quite young in planetary terms. Only two impact craters have been found, far less than would be expected in comparison with other Saturn satellites. This indicates that craters must be destroyed by recently or currently active geologic processes. The major geologic processes shaping all planetary surfaces are volcanism, tectonism, impact cratering and erosion. Since the scars of impacts are being erased on a large scale, one or more of the other processes must be responsible. The first radar images of Titan (October 2004) showed evidence of volcanism (Fig. 5) in the form of a dome and extensive flows. Ridges that may be tectonic in nature were shown in radar images obtained a year later. In the meantime, other radar images showed alluvial deposits (Fig. 6) and fields of dunes (Fig. 7), indicating erosion by liquids and wind on a large scale. Titan appears to be a dynamic world, in some ways remarkably Earth-like in its geology. Few well-preserved impact craters are seen on Earth, because our planet's active geology has erased these features formed relatively early in the history of the Solar System. In general, the older a surface has remained unmodified, the more craters it will display. Therefore, our own Moon, a largely dead world, shows a plethora of impact scars. In comparison, Titan looks far from dead.

However, when looked at in detail, Titan and Earth are not at all similar. While water carves river valleys on Earth, the liquid on Titan is likely to be methane. Liquid methane may exist in lakes or pools, such as in some dark areas imaged by Cassini. Methane may also come down in the form of rain and we know from Cassini orbiter images that methane is present in clouds near the south pole. Rivers of methane may carve dendritic channels such as those seen in radar images. Volcanism on Titan is also nothing like Earth's. The flows and dome seen in radar images were formed by cryovolcanism (4, 11), where the "magma" is not molten rock but liquid water coming from below the frozen surface, most likely mixed with other components such as ammonia.

One of the most remarkable surprises from the orbiter's radar images was the discovery of "sand seas" (16), that is, large areas of the surface covered by dunes that appear similar to longitudinal dunes on Earth, such as those in Namibia. This is further proof that Titan's surface and atmosphere are indeed dynamic and that weather on Titan has a great influence on the appearance of the surface.

Above: *Fig. 5: Synthetic aperture radar image of a possible cryovolcanic flow on the surface of Titan, where water-rich liquid likely welled up from Titan's interior. The image was acquired on 26 October 2004, when the Cassini spacecraft flew approximately 2,500km above the surface and acquired radar data for the first time. The radar illumination was from the south: dark regions may represent areas that are smooth, made of radar-absorbing materials, or are sloped away from the direction of illumination. The bright flow-like feature stretches from upper left to lower right across this image, with connected 'arms' to the east. The fact that the lower (southern) edges of the features are brighter is consistent with the structure being raised above the relatively featureless darker background. The image covers an area about 150km^2. Image courtesy NASA/JPL.*

Above right: *Fig. 6: The Cassini radar system imaged this area during the spacecraft's third close flyby of Titan on 15 February 2005. The bright lines are interpreted as channels in which fluid flowed toward the bright area in the upper right. Areas that appear bright at radar wavelengths may be rough or inclined toward the direction of illumination. The bright area in this image could have received outflows of debris from the channels, making the surface appear radar bright. In this sense, the area may resemble the rubble strewn plains in the region where the Huygens probe landed (see Fig. 9). The fluid carrying the debris was most likely liquid methane. The longest channel in the feature is approximately 200km long. The seams running across the image are an effect of the matching of the different radar beams to assemble the full image. Image courtesy NASA/JPL.*

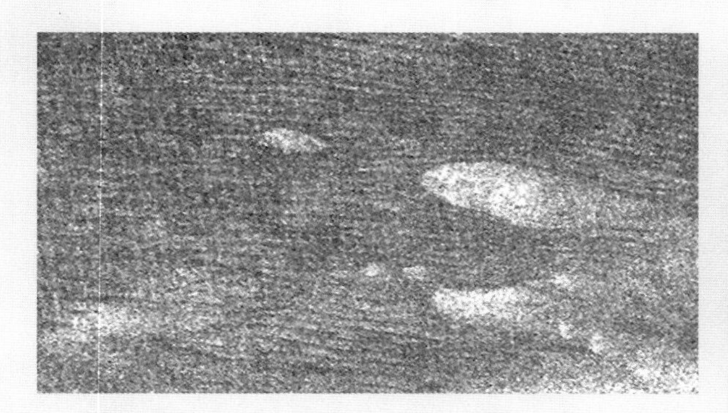

Huygens landing

The wok-shaped Huygens probe carried six instruments to study Titan's surface and atmosphere in-situ (8, 17). During the 2.5hr descent, instruments made measurements of atmospheric conditions, including the direction and speed of winds. At 120km altitude, Titan's winds, blowing mostly in the direction Titan is rotating, reached 120m/s, which is faster than Titan rotates. Titan's atmosphere is therefore referred to as "super-rotating". Although this had been predicted (18), Cassini observations confirmed the prediction. In contrast, winds at the surface were very weak, about 1m/s. Can these light winds account for the formation of dunes on the surface, or are there stronger winds at times? The question remains open.

The composition of the atmosphere was also analyzed during descent and preliminary results indicate the presence of organic compounds containing nitrogen, which may include amino, imino and nitrale compounds. These aerosols are thought to fall steadily on the surface as "organic rain", depositing a global layer that may be as thick as 1km. This

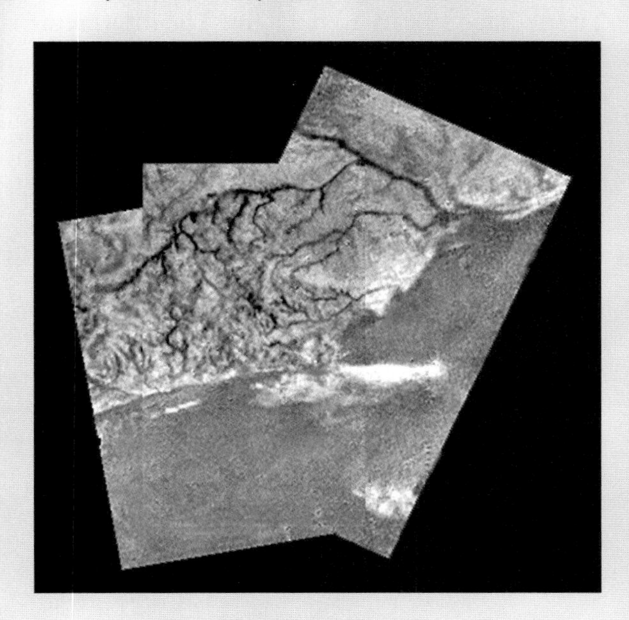

■ ***Above left:*** *Fig. 7: Large areas of this Cassini synthetic aperture radar image of Titan are covered with long, dark ridges spaced about 1km to 2km apart. They curve slightly around teardrop-shaped bright terrain, giving the impression of a Japanese garden of sand raked around boulders. The bright material appears to be high-standing rough material that the ridges bend around. This suggests that the ridges are dunes that winds have blown across the surface of Titan from left to right (roughly west to east). This image was taken during the ninth Titan flyby on 28 October 2005, at a distance of about 1,300km. The image covers an area roughly 140km by 200km. Image courtesy NASA/JPL.*

■ ***Above:*** *Fig. 9: Huygens view of Titan's surface obtained using the Descent Imager/Spectral Radiometer on 14 January 2005. This is a coloured view, following processing to add reflection spectra data and gives a good indication of the actual colour of the surface. Several rounded pebbles are seen. The two just below the middle of the image are about 15cm (left) and 4cm (center) across, at a distance of about 85 centimetres from Huygens. The surface is darker than originally expected, consisting of a mixture of water and hydrocarbon ice. There is also evidence of erosion at the base of these objects, indicating possible fluvial activity. Image courtesy NASA/JPL/ESA/ University of Arizona.*

■ ***Opposite:*** *Fig. 8: This mosaic of three frames from the Huygens Descent Imager/ Spectral Radiometer (DISR) instrument was obtained on 14 January 2005. The image shows a complex dendritic network of drainage channels, possibly cut by liquid methane. The bright area that the channel cuts is interpreted as a high ridge. The channels are possibly flowing down into the darker area at the bottom of the image mosaic, interpreted as a river or lake bed. Image courtesy NASA/JPL/ESA/University of Arizona.*

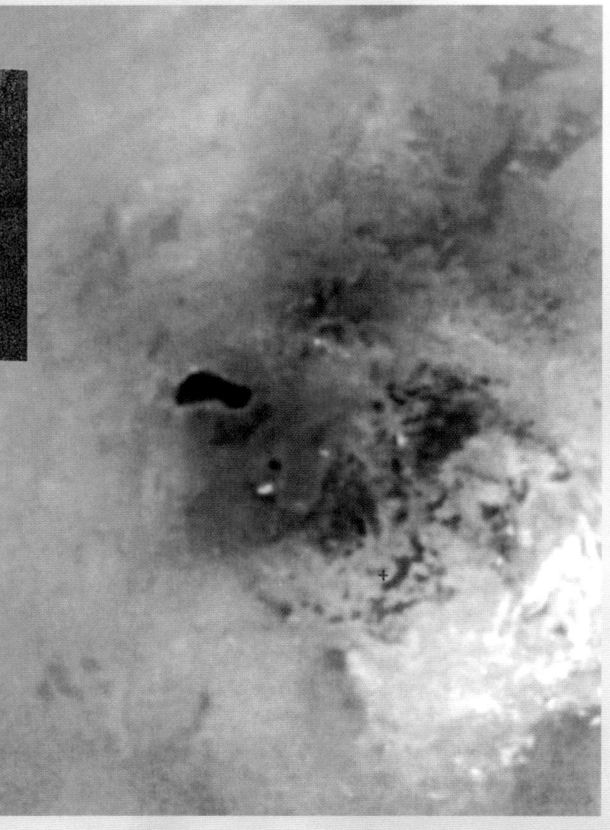

process removes atmospheric methane and it has long been known that Titan's methane must be re-supplied either continuously or episodically. Volcanic activity has been proposed as a mechanism (10).

Huygens' instruments confirmed that methane on Titan is involved in a phase change cycle similar to water on Earth (17). Both the Descent Imager/Spectral Radiometer (DISR) and the Gas Chromatograph Mass Spectrometer (GCMS) detected methane haze in the lower regions of the atmosphere (19, 20), and the Surface Science Package (21) found that the ground at the landing site had the consistency of damp sand. This can probably be explained in terms of a soil that contains ice chips, precipitated aerosols and liquid methane.

The nearly flawless operation of Huygens ensured that images of Titan's surface were obtained both during descent and after landing. DISR images (20) revealed channel networks with dendritic patterns (Fig. 8) that carved light-coloured terrain and appear to empty out onto darker coloured terrain with the appearance of a river or lake bed. Images taken after landing are like those of a dry riverbed on Earth, showing rounded boulders mostly 5-15cm in diameter lying above a darker, finer-grained substrate (Fig. 9). The composition of the boulders is unknown, but they are likely ice, coated with hydrocarbons. In colour, the view is orange, because of the much greater attenuation of blue light by Titan's haze relative to red light. The interpretation of the area seen during descent is that the geologic features result from the flow of low-viscosity fluids driven by variations in topography. Dendritic networks are thought to be the result of precipitation, while more stubby networks may have been spring-fed. Surprisingly, given the evidence from both orbiter and lander that Titan's surface has been significantly shaped by the flow of liquids, so far there is no evidence of liquids on the surface. Tantalizing images of a possible shoreline (Fig. 10) and lake (Fig. 11) have been obtained, but so far we have no conclusive evidence that these and other areas are filled with liquid hydrocarbons. This is one of the many mysteries that the continuing Cassini mission will try to unravel.

■ **Above:** *Fig. 10: Possible shoreline on Titan's surface imaged by the Cassini radar on 7 September 2005. The bright, rough region on the left side of the image seems to be topographically high terrain that is cut by channels and bays. The boundary of the bright (rough) region and the dark (smooth) region appears to be a shoreline. The patterns in the dark area indicate that it may once have been flooded, with the liquid having at least partially receded. The image is 175 km high and 330 km wide. Image courtesy NASA/JPL.*

■ **Above:** *Fig. 11: This view of Titan's south polar region reveals an intriguing dark feature that may be the site of a past or present lake of liquid hydrocarbons. The true nature of this feature, seen here at left of center, is not yet known, but the shore-like smoothness of its perimeter and its presence in an area where frequent convective storm clouds have been observed by Cassini and Earth-based astronomers are consistent with its being an open body of liquid on Titan. If this interpretation is correct, then other very dark but smaller features seen in the south polar region, some of which are captured in this image, may also be the sites of liquid hydrocarbon reservoirs. A red cross below center of the image marks the south pole location. The brightest features seen here are methane clouds. This view is a composite of three Cassini spacecraft narrow-angle camera images taken over several minutes during Cassini's distant flyby on 6 June 2005, from a distance of about 450,000km from Titan. Image courtesy NASA/JPL/Space Science Institute.*

References

1. Coustenis, A., and Taylor, F.W. (1999). *Titan: the Earth-like Moon.* (World Scientific)

2. Lorenz, R. and Mitton, J. (2002). *Lifting Titan's Veil* (Cambridge University Press)

3. Russell, C., and chapters therin (2002, 2004). *The Cassini Huygens Mission*, volumes 1, 2, and 3. Kluwer Academic Publishers, Dordrecht/Boston/London.

4. Elachi, C., et al. (2005). Cassini's RADAR first views of the surface of Titan. Science, 308, 970-974.

5. Porco, C.C., et al. (2005). Imaging of Titan from the Cassini spacecraft. Nature, 434, 159-168.

6. Sotin, C., et al. (2005). Release of volatiles from a possible cryovolcano from near-infrared imaging of Titan. Nature 435, 786-789.

7. Harland, D.M. (2002). *Mission to Saturn: Cassini and the Huygens Probe.* (Springer-Praxis)

8. Lebreton, J-P., et al. (2005). An overview of the descent and landing of the Huygens probe on Titan. Nature, 438, 758-764.

9. Lorenz, R., and the Cassini RADAR Team (2005). Titan's methane monsoon: Evidence of catastrophic hydrology from Cassini RADAR. Bulletin of the American Astronomical Society, 37 (3), abstract # 53.07.

10. Lunine, J., et al. (1983) Ethane ocean on Titan. Science 222, 1229-1230.

11. Lopes, R., et al. (2006). *Cryovolcanic features on Titan's surface as revealed by the Cassini Titan Radar Mapper.* Submitted to Icarus.

12. Waite, J.H. et al. (2005), Ion neutral mass spectrometer results from the first fly-by of Titan. Science, 308, 982-986.

13. Krimigis, S.M., et al. Dynamics of Saturn's Magnetosphere from MIMI during Cassini's orbital insertion. Science, 307, 1270-1273.

14. Teaby, N. Cassini at Titan. Astronomy & Geophysics, 46, 5.20-5.25.

15. Elachi, C., et al. (2006). *Titan Radar Mapper Observations from Cassini's Ta and T3 Fly-bys.* Submitted to Nature.`

16. Lorenz, R., et al., *The Sand Seas of Titan: Cassini RADAR observations of Equatorial Fields of Longitudinal Dunes.* Submitted to Science.

17. Owen, T. (2005) Huygens Rediscovers Titan. Nature, 438, 756-757.

18. Del Genio, A.D. et al. (1993). Equatorial superrotation in a slowly rotating GCM: Implications for Titan and Venus. Icarus, 101, 1-17.

19. Niemann, H.B. et al. (2005). The abundances of constituents of Titan's atmosphere from the GCMS instrument on the Huygens probe. Nature, 438, 779-784.

20. Tomasko, M.G. et al. (2005). Rain, winds, and haze during the Huygens probe's descent to Titan's surface. Nature, 438, 765-778.

21. Zarnecki, J.C. et al. (2005). A soft solid surface on Titan as revealed by the Huygens Surface Science Package. Nature, 438, 792-795.

Rosaly Lopes *is Lead Scientist for Geophysics and Planetary Geosciences at NASA's Jet Propulsion Laboratory. She is also the Investigation Scientist for the RADAR instrument on Cassini.*

4.6 Catching Fragments OF OUR PAST...

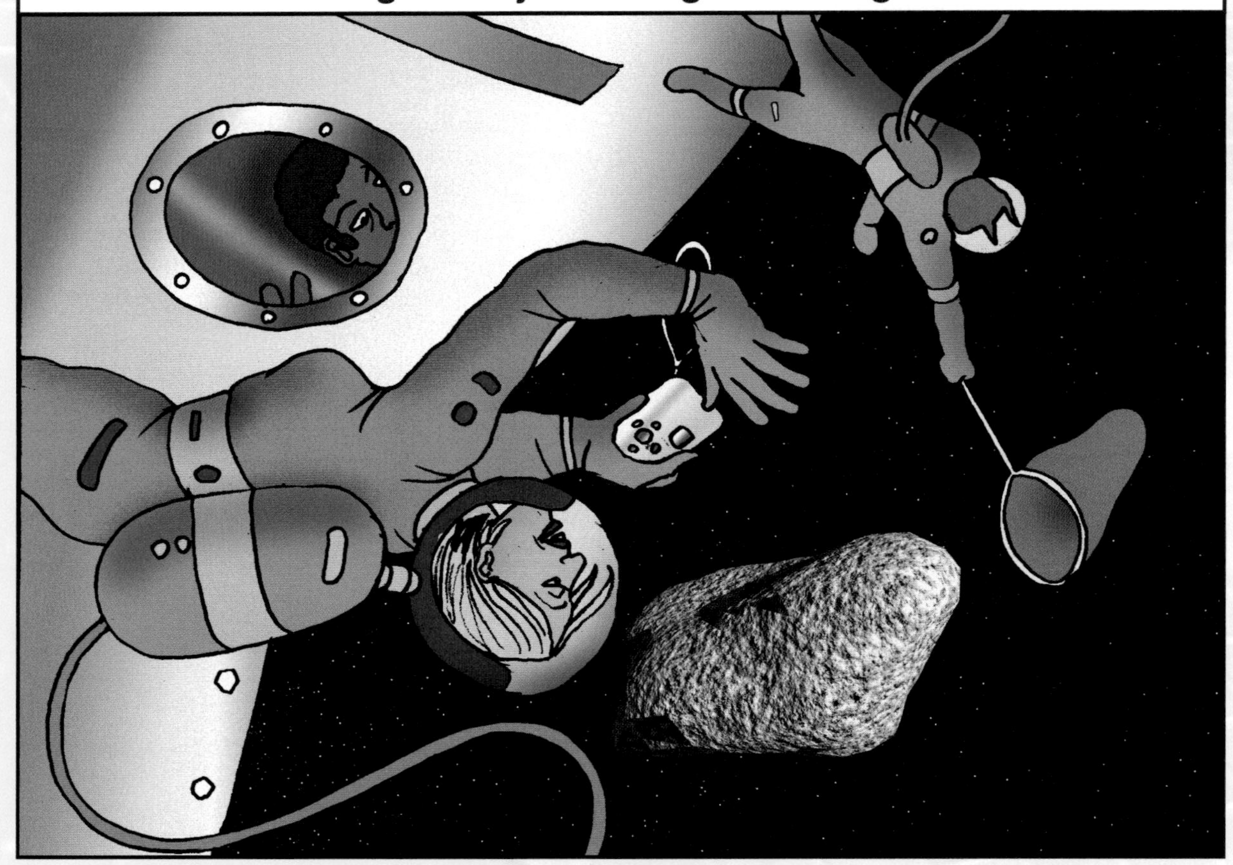

In 2005, the Japanese space probe Hayabusa made a spectacular interception of asteroid Itokawa, while in January 2006 the American space agency NASA brought back to the Utah desert on Earth the Stardust space probe sample return capsule after a lengthy mission to intercept a comet and collect samples of interstellar and cometary dust.

Here, *John Mason* explains why asteroids and comets are the latest target for space exploration and what has been learned so far.

Intercepting COMETS and ASTEROIDS

DURING THE first three decades of the Space Age, the main emphasis of Solar System exploration had been on the Moon and planets and the major moons of those planets. But there is more to the Sun's family than these main players. Orbiting between the planets are countless minor bodies – the rocky asteroids and icy comets. Planetary scientists gradually came to realize that a complete understanding of the formation and evolution of our planetary system would only be obtained if they were able to study these smaller Solar System bodies in detail.

Comets have been known since ancient times. These ghostly visitors appeared unexpectedly in the sky at irregular intervals, moving against the background stars from night to night, often striking great fear into those who saw them. So-called Great Comets were undoubtedly bright, some being visible in twilight or broad daylight and sporting magnificent tails which stretched across the sky. Today we know that comets are icy visitors, originating in the twilight zone at the edge of the Solar System beyond the major planets. They are thought to have been around, relatively unchanged, since the early days of the Solar System's formation about 4.5bn years ago. We believe they contain the most primitive materials in the Solar System and this makes them 'time capsules' of valuable information for scientists interested in learning about the early history of the Sun and planets. Comets are also thought to have provided much of the water that covered a dry early Earth with oceans, making life possible. Some scientists have even suggested that comets might have seeded the building blocks of life on our planet around 3.8bn years ago.

Asteroids have been known since 1 January 1801, when the first and largest asteroid, Ceres, was discovered. Since that time several hundred thousand asteroids have been found, most of them within the so-called Main Belt, which extends from just beyond the orbit of Mars out to some way inside the orbit of Jupiter. In the early history of the Solar System, it was Jupiter's gravity that continually stirred the rocky debris of the asteroid belt and prevented the material clumping together to form a planet. The scientific interest in asteroids is due largely to their status as the debris left over from the processes that formed the inner rocky planets, including Earth. Asteroids are also the sources of most meteorites that have fallen to the Earth's surface and if certain as-

Few sights are more impressive than a bright comet hanging in the sky just after sunset or shortly before dawn. This view of Great Comet Hale-Bopp was taken on 9 March 1997 at 0500 hrs GMT. The comet's two principal tails are clearly shown. The straight bluish tail pointing directly away from the Sun is the plasma or ion tail. The broad, curving yellowish tail is composed of microscopic dust grains. Image courtesy of Glyn Marsh.

teroids can be identified as the sources of some of the most well-studied meteorites, this will provide important information on the chemical mixture and conditions from which the inner planets formed. Asteroids also offer a potential rich source of minerals that could be exploited for the future exploration and colonization of our Solar System.

The orbits of some asteroids have been perturbed well beyond the Main Belt and a number of these have orbits which approach that of the Earth or even cross it, so there is the possibility that a few of these may collide with our planet. Indeed, both comets and asteroids may hit the Earth from time to time. Such events have happened in the past and will undoubtedly occur again in the future. There is thus considerable interest in finding out more about these bodies which have the potential to cause great destruction on a global scale, drastically altering the biosphere and even triggering mass extinctions. The chance of an asteroid or comet striking the Earth is remote, but the devastating consequences of such an impact suggests we should closely study these bodies to understand their sizes, compositions, structures and future trajectories.

Prior to the first close-range views from spacecraft, there was no complete agreement as to the precise nature of comets. Harvard University astronomer Fred Whipple had proposed the idea that all cometary activity stemmed from a tiny nucleus consisting mostly of frozen water ice with plenty of dirt mixed in – the so-called "dirty snowball" model. At great distances from the Sun, the nucleus would be frozen, inert and generally invisible. As it approached the Sun, the surface of the nucleus would be warmed and its ices would sublimate (turn from solid ice into a vapour), releasing dust which together with the cometary gas would produce the fuzzy head or coma and the tails of gas and dust. Other scientists were of the opinion that comets were just an orbiting swarm of dust particles, the swarm being fairly compact when closest to the Sun, but much more extended when furthest away – the so-called "flying gravel bank" model. Such a major disagreement would only be finally resolved when close-range images of comets could be obtained from spacecraft.

Above right: *The nucleus of comet Halley, composed of 68 images (of varying resolution) acquired by Giotto's Halley Multicolour Camera. The night side of the elongated nucleus is silhouetted against a background of bright dust. Jets of gas and dust can be seen originating from three regions on the nucleus. The bright area within the night-side of the nucleus is probably a hill about 500m high. Image courtesy of the Max-Planck Institut für Aeronomie, Katlenburg-Lindau, Germany.*

Targeting Halley's comet

Of all the known comets, Halley's comet is undoubtedly the most famous. It is the only bright, highly active comet with a well-known orbit, returning to the inner Solar System every 76 years or so. Because of its high activity, it has been recorded at every appearance since that of 240 BC. The orbit of Halley's comet is inclined at 18° to the ecliptic (the plane of Earth's orbit around the Sun). At its last return in 1985/86, the first during the Space Age, the comet crossed the ecliptic, travelling north, on 9 November 1985 and again, travelling south, on 10 March 1986. The second of these crossings was only four weeks after the comet's perihelion passage (when it was closest to the Sun), with the comet still only 126m km from the Sun and highly active. To send a spacecraft to this point required one of the lowest launch energies of all possible cometary missions, so that even with modest launch vehicles a substantial scientific payload could be carried to the comet. It is therefore no surprise that six spacecraft from four space agencies were despatched to intercept the comet at this time, taking advantage of what would be a 'once-in-a-lifetime' opportunity.

There was however one disadvantage in choosing Halley's comet as a target. The comet's orbit is retrograde, which means that it travels around the Sun in the opposite direction to the Earth – and spacecraft launched from the Earth. Consequently, the spacecraft would meet the comet more or less 'head-on', leading to a very high relative flyby velocity of around 70km/sec – fifty times faster than a rifle bullet. Spacecraft would be in the vicinity of the comet for only a few hours, so scientific data would have to be transmitted at a very high rate. In addition, dust from the comet would hit the spacecraft at high velocity so some form of shielding would be required for any spacecraft flying through the comet's inner coma.

Onward to Halley

The US space agency NASA found itself in the embarrassing position of not being able to send a dedicated spacecraft to Halley's comet because of budgetary cutbacks. However, one crumb of comfort came with the approval of a probe to intercept the relatively inactive short-period comet Giacobini-Zinner in September 1985, but even this involved utilising an existing spacecraft ISEE-3 (International-Sun-Earth-Explorer-3). ISEE-3 had been launched in 1978 and was in orbit around the Earth monitoring the activity of the Sun and the effects of solar activity on Earth's outer atmosphere. Through a series of complex manoeuvres and orbital changes, involving five swing-by passages of the Moon, ISEE-3 gained enough energy to put it into a path that would take it to Giacobini-Zinner. The probe, now renamed ICE (International Comet Explorer), passed through the tail of the comet on 11 September 1985, swooping just 3,000km from the heart of the comet, but was unable to confirm the presence of a nucleus as there was no camera on board. Nevertheless it provided the first ever in situ measurements of a comet and its environment, in particular the way in which the plasma and embedded magnetic field of the solar wind (a stream of electrically-charged particles from the Sun) interacted with the gas in the comet's coma and tail. At the end of March 1986, ICE passed comet Halley at a distance of 28m km on the sunward side and monitored how the solar wind was affected by the comet.

The main hopes for investigating Halley's comet from space in 1986 rested with three projects: the Japanese probes Sakigake and Suisei, Russia's Vega-1 and Vega-2 craft and Europe's Giotto. The two Japanese craft were almost identical except for the scientific experiments they carried. Sakigake (Japanese for 'forerunner') carried three experiments, a plasma wave probe, a solar wind experiment and a magnetometer. Suisei (Japanese for 'comet') carried two experiments, an ultraviolet imager and a plasma experiment for the observation of solar wind plasma and cometary ions. Sakigake passed the comet at a distance of 7m km on the sunward side while Suisei passed by at a distance of 151,000km. Ultraviolet images of the comet's vast hydrogen coma were recorded by Suisei from November 1985 through to April 1986 and enabled the comet's water production rate to be determined throughout the period of observation.

The nucleus at last!

The two Russian Vega craft took advantage of a unique opportunity to combine missions to Venus and Halley's comet, each spacecraft comprising both a Venus lander and a Halley flyby probe. Indeed the name Vega was a contraction of the Russian words Venera (Venus) and Gallei (Halley). Having successfully dropped their Venus landers (and the first balloons into the Venusian atmosphere) en route, Vega-1 and Vega-2 encountered Halley's comet on 6 and 9 March 1986, respectively. Each spacecraft carried a payload of 14 scientific experiments, including a TV system for imaging the inner coma and cometary nucleus. Successful imaging from the Vega spacecraft required a steerable platform which could be automatically pointed with great accuracy at the nucleus.

The imaging systems on both Vega spacecraft worked for 9-10 hours, producing about 1,500 images, and for a roughly 20min period during this time, before and after closest approach, the cometary nucleus was resolved at last. Fred Whipple had been right! At closest approach, Vega-1 passed just 8,890km from the nucleus and Vega-2 slightly closer at 8,030km. The images enabled a 3D model of the comet's nucleus to be derived and this showed it to be a highly elongated, roughly peanut-shaped body, measuring about 15 x 8km. The nucleus was also found to be extremely dark, reflecting just 4% of the light falling upon it.

The Vega results were confirmed in spectacular fashion by Europe's Giotto probe which hurtled thorough the inner coma of the comet, passing just under 600km from the nucleus on 13/14 March 1986. At this almost 'suicidal' flyby distance, it was essential to protect the craft from ultra-high-velocity dust particle impacts, so Giotto carried a dual-sheet bumper shield, the front sheet of which faced forwards during the flyby. Giotto carried 10 hardware experiments mounted on a platform behind the rear sheet of the bumper shield. An additional radio science experiment showed how the spacecraft was decelerated by dust particle impacts as it traversed the coma. Most important was the Halley Multicolour Camera (HMC) used for imaging the nucleus. This was mounted behind the spacecraft bumper shield and so protected from direct dust-particle impacts. A 45° deflecting mirror was used to look forwards at the comet and there was a long cylindrical baffle made of Kevlar to protect the HMC optics from stray light. Altogether the HMC took more than 2,300 images in about three hours and the nucleus was visible in several hundred of these images at distances as far as 100,000km. The last useful image was obtained from a distance of 1,675km, a few seconds before closest approach when communications with the spacecraft were interrupted because of dust particle impacts.

The HMC images confirmed that the nucleus of comet Halley is an irregular, elongated, slowly rotating body measuring 15.3 x 7.2 x 7.2km. The best estimate for the density was around 0.6 gm/cm^3. The rotation of the nucleus turned out to be far more complex than had been imagined; the nucleus appears to spin about its long axis in a period of 7.4 days, while precessing (gyrating) about the short axis in just 2.2 days. Three major jets of dust and gas were evident, spurting out through fissures in the very dark, dusty crust of the nucleus. However, this dust and gas emission was restricted to only 10-15% of the surface area. A wide range of so-called CHON particles, containing significant amounts of the elements carbon, hydrogen, oxygen and nitrogen, were detected leaving the nucleus. It is estimated that about 6m depth of material is lost from an active region on each orbit around the Sun.

Other comets than Halley

After the highly successful close-range flyby of comet Halley, it was found that most of Giotto's main instruments were still functioning, with the important exception of the HMC whose deflecting mirror and baffle had clearly been irreparably damaged by dust particle impacts shortly before closest approach. The spacecraft was redirected towards Earth before it was put into 'hibernation mode'. Giotto eventually swung by the Earth on 2 July 1990 and taking advantage of a gravity assist it was redirected towards the short-period comet Grigg-Skjellerup. This is an old, relatively inactive comet, whose orbit is well known, but it provided an interesting contrast to Halley. During this Giotto Extended Mission (GEM), the spacecraft successfully encountered comet Grigg-Skjellerup on 10 July 1992. The closest approach distance was about 200km. The payload was switched on the evening before encounter, and eight experiments were operated, yielding a surprising wealth of interesting data. One instrument picked up the first presence of cometary ions 600,000km from the nucleus, about 12hr before the closest approach, and the impacts of fairly large dust particles were also detected. Following the Grigg-Skjellerup encounter, Giotto operations were terminated on 23 July 1992.

After the comet Halley encounter in 1986, scientists had to wait over 15 years for their next close-up views of a cometary nucleus. They came courtesy of a spacecraft called Deep Space 1, launched from Cape Canaveral on 24 October 1998. The objectives of the spacecraft's primary mission were the testing of 12 advanced technologies with the potential to lower the costs and risks of future space missions. Having succeeded in these tasks, Deep Space 1 embarked on an extended mission, which took it first past the asteroid 9969 Braille on 29 July 1999, although because of a software crash shortly before closest approach the pictures were disappointingly fuzzy. The craft then went on to make a flyby of periodic comet Borrelly on 22 September 2001, passing just 2,200km from the nucleus, returning detailed images along with other science data.

Borrelly's nucleus turned out to be roughly half the size of Halley, and even more elongated, measuring 8.0 x 3.2 x 3.2km. Its shape was reminiscent of a tenpin bowling pin. Borrelly was less active than Halley with only one main dust and gas jet visible, although this may have been produced from three discrete active areas on the nucleus. Once again only a small percentage of the surface area was active. Fortunately, the main jet did not appear to point towards the spacecraft,

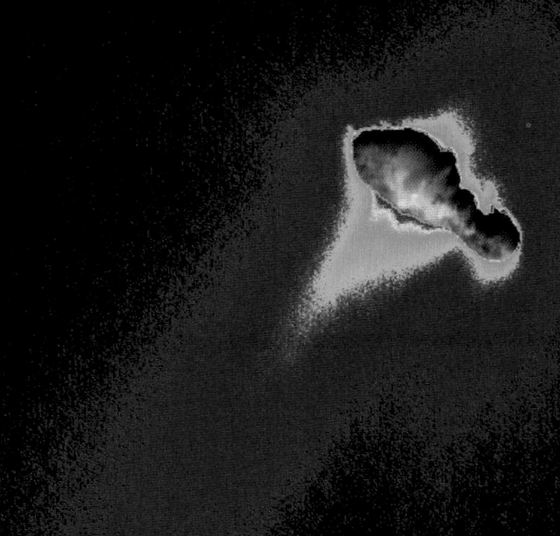

This composite of images from NASA's Deep Space 1 spacecraft shows comet Borrelly's nucleus. (An enlarged view of the bowling-pin shaped nucleus is shown in the inset.) False colour is used to reveal details of dust jets escaping from the nucleus and the cloud-like coma of dust and gases surrounding it. The Sun shines from the bottom of the image. The red bumps near the nucleus show where the jet resolves into three distinct, narrow jets, which probably come from discrete sources on the surface. Image courtesy of NASA Jet Propulsion Laboratory.

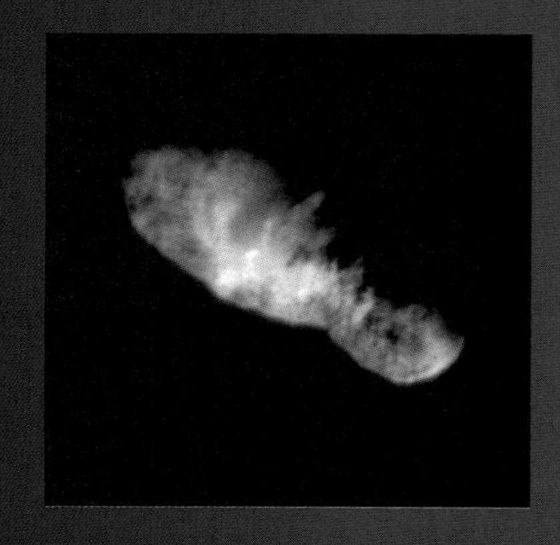

because it had no dust shields and yet it survived its passage through the dusty coma intact. Deep Space 1 was retired on 18 December 2001.

The first comet samples

Even while Deep Space 1 was still on its way to encounter comet Borrelly, NASA launched an even more ambitious mission with the objective of capturing samples of interstellar and comet dust and returning them to Earth - the first comet sample return. Launched on 7 February 1999, the Stardust spacecraft began a 7-year journey that would take it on three orbital loops around the Sun. The first loop was a 2-year orbit, with a trajectory correction manoeuvre near aphelion to set up an Earth gravity assist on 15 January 2001. This expanded the orbit into a longer 2.5-year loop, which the spacecraft flew twice, bringing it back to Earth once again on 15 January 2006. These longer orbits provided adequate time for a comprehensive collection of interstellar dust, the dust particle collections being carried out during February-May 2000 and August-December 2002. The minute dust particles were collected in a microporous silica aerogel, contained in compartments within a two-sided, tennis-racket-shaped array, that deployed from the clamshell-like sample return capsule. One side of the array was used to collect the samples of interstellar dust, while the reverse side would be used during the comet encounter.

Further trajectory correction manoeuvres in January 2002 and July 2003 set up a flyby of the short-period comet Wild-2 (pronounced "Vilt-2") on 2 January 2004, five years after launch. On 2 November 2002, Stardust had passed within about 3,300km of the Main Belt asteroid 5535 Annefrank. The asteroid flyby was used as an engineering test of the ground and spacecaft operations to be implemented during the comet encounter. Wild-2 was selected as the target because it is a recent arrival into the inner Solar System, having been perturbed into its present orbit after a close encounter with giant Jupiter in 1974. Consequently, as a comet new to our neighbourhood, the surface of the nucleus should be relatively unaffected by repeated passages close to the Sun. Indeed, Wild-2 had made only five trips around the Sun in its new, closer orbit by the time Stardust arrived to examine it. As it closed in on comet Wild-2, Stardust endured an intense bombardment by dust particles surrounding the nucleus, but an arrangement of bumper shields protected both the solar panels and the main spacecraft body.

The spacecraft passed just 236km from the nucleus of comet Wild-2 – the closest encounter at that time – and acquired 72 images. There were plenty of surprises. The nucleus was less elongated, more round and smaller than either Halley or Borrelly, measuring 5.5 x 4.0 x 3.3km. The nucleus was pockmarked by several roughly circular depressions, with flat floors and surprisingly steep walls. Although the gravity on a body as tiny as Wild-2 is extremely low, such steep slopes suggest the materials that form the nucleus have enough internal strength to hold together despite gravity's tendency to flatten out such slopes. In general, the surface of the nucleus consists of gently undulating hills with valleys in between, the high spots being roughly 100-200 metres above the low spots.

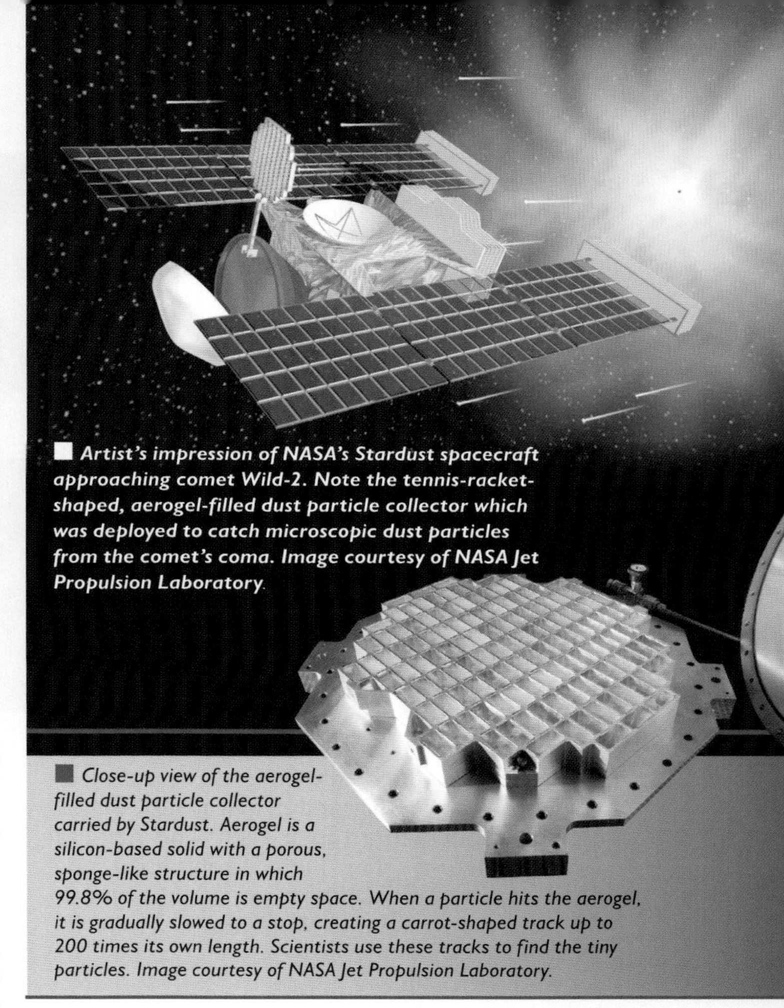

Artist's impression of NASA's Stardust spacecraft approaching comet Wild-2. Note the tennis-racket-shaped, aerogel-filled dust particle collector which was deployed to catch microscopic dust particles from the comet's coma. Image courtesy of NASA Jet Propulsion Laboratory.

Close-up view of the aerogel-filled dust particle collector carried by Stardust. Aerogel is a silicon-based solid with a porous, sponge-like structure in which 99.8% of the volume is empty space. When a particle hits the aerogel, it is gradually slowed to a stop, creating a carrot-shaped track up to 200 times its own length. Scientists use these tracks to find the tiny particles. Image courtesy of NASA Jet Propulsion Laboratory.

Prior to the Stardust encounter, it was assumed that the coma had a dust particle density that increased uniformly toward the nucleus, but Stardust ran into three distinct 'sheets' of dust as it approached the comet. During the flyby, Stardust's aerogel-filled dust particle collector was used to sample dust particles from the comet's coma for comparison with the interstellar dust particles collected earlier. After exposure, the array folded compactly for stowage inside the sample return capsule for the trip back to Earth. This dramatic event took place on 15 January 2006, seven years after launch, when the capsule successfully made a soft landing at the U.S. Air Force Utah Test and Training Range. The microscopic particles of interstellar and comet dust collected by Stardust were then taken to the planetary material curatorial facility at NASA's Johnson Space Center, Houston, Texas, for analysis. Here the cometary dust particles are being extracted from the aerogel and analysed to determine their detailed composition, in a study that could take years.

While Stardust was still completing its second orbital loop of the Sun, NASA's Comet Nucleus Tour (CONTOUR) spacecraft – its second mission dedicated solely to exploring comets – blasted off on 3 July 2002. CONTOUR's orbit was designed to loop around the Sun and back to Earth for annual gravity assists toward its targets. Such manoeuvres would alter CONTOUR's trajectory and help it reach several comets without using much fuel. The flexible four-year mission plan included planned encounters with comets Encke (on 12 November 2003) and Schwassmann-Wachmann 3 (on 19 June 2006), although it was possible to add an encounter with a 'new' and scientifically valuable comet coming in from the outer Solar System, should a suitable one be discovered. It was hoped that CONTOUR would fly closer to each cometary nucleus than ever before (within 100km), gathering detailed, comparative data on these dynamic objects and obtaining the first hard evidence of comet nuclei composition. Following its successful launch, the CONTOUR spacecraft orbited Earth until 15 August 2002. Then disaster struck! Some major malfunction, most probably while CONTOUR was firing its main solid-propellant motor to enter its comet-chasing orbit around the Sun, caused the spacecraft to break in several pieces. Despite many attempts to make contact with the craft, ground controllers reluctantly had to face up to the fact that the CONTOUR mission was lost.

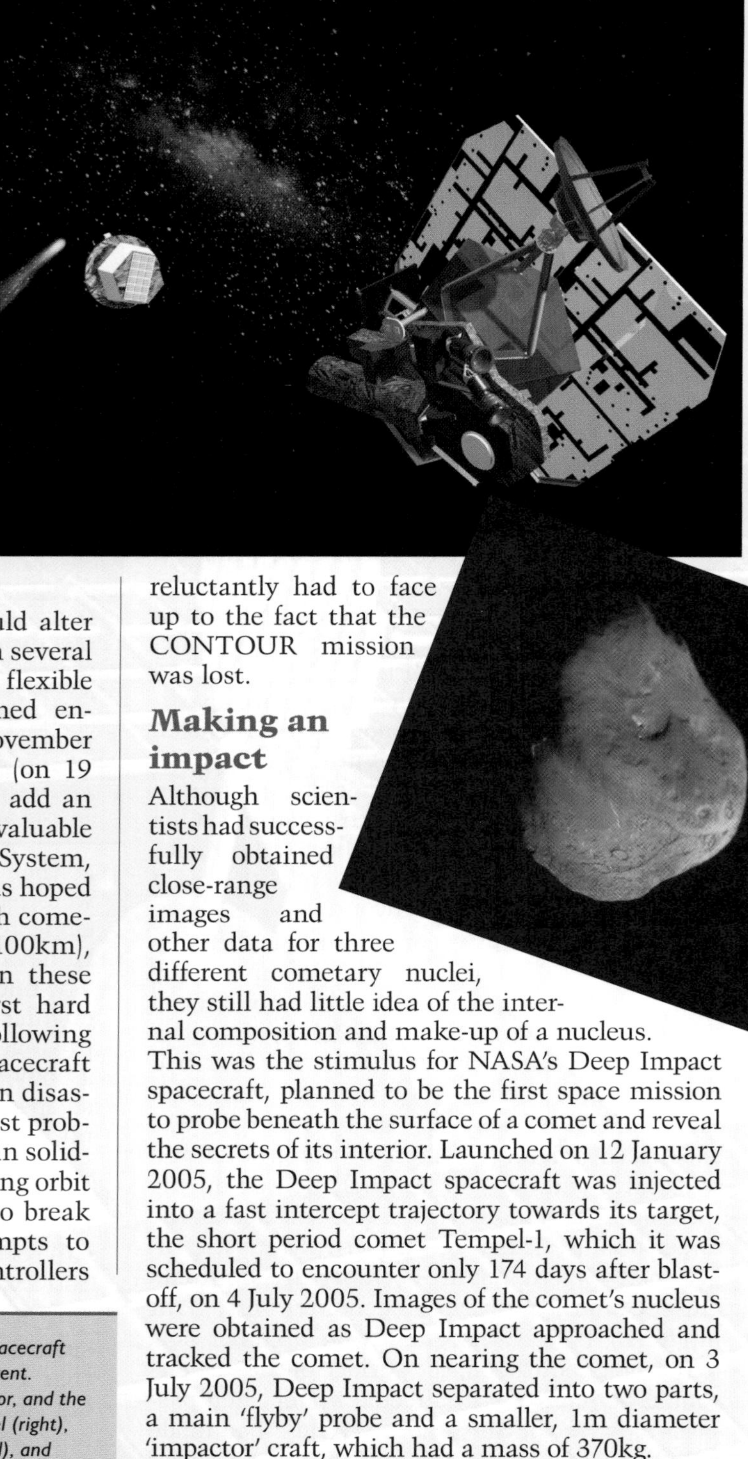

Above: *Artist's impression of the Deep Impact flyby spacecraft releasing the 370kg impactor the day before the impact event. Pictured from left to right are comet Tempel-1, the impactor, and the flyby spacecraft. The flyby spacecraft includes a solar panel (right), a high-gain antenna (top), a debris shield (left, background), and science instruments for high and medium resolution imaging, infrared spectroscopy, and optical navigation (yellow box and cylinder, lower left). Image courtesy of NASA/JPL-Caltech/University of Maryland.*

Above right: *A composite image of the nucleus of comet Tempel-1 from Deep Impact's High Resolution Imager. Each image at closer range replaced equivalent locations observed at a greater distance. Note the unusual, extremely smooth regions, one lower left, and the other upper right. The impact site was just above the lower of the two circular crater-like features towards the bottom of the nucleus. Image courtesy of NASA/JPL-Caltech/University of Maryland.*

Making an impact

Although scientists had successfully obtained close-range images and other data for three different cometary nuclei, they still had little idea of the internal composition and make-up of a nucleus. This was the stimulus for NASA's Deep Impact spacecraft, planned to be the first space mission to probe beneath the surface of a comet and reveal the secrets of its interior. Launched on 12 January 2005, the Deep Impact spacecraft was injected into a fast intercept trajectory towards its target, the short period comet Tempel-1, which it was scheduled to encounter only 174 days after blast-off, on 4 July 2005. Images of the comet's nucleus were obtained as Deep Impact approached and tracked the comet. On nearing the comet, on 3 July 2005, Deep Impact separated into two parts, a main 'flyby' probe and a smaller, 1m diameter 'impactor' craft, which had a mass of 370kg.

The self-guided impactor used its thrusters to move into the path of the oncoming comet, hitting it 24hr later, on the sunlit side, at a relative speed of 10.3km/sec (37,000km/hr). A camera on the impactor captured and relayed images of the comet's nucleus seconds before the collision. With such a high impact velocity, the energy released was equivalent to exploding 4.5 tonnes of

TNT. As expected, the impact blasted a crater, perhaps as large as a football stadium, in the comet, ejecting a plume of ice and dust from the crater and revealing fresh material beneath. After releasing the impactor, the main flyby spacecraft manoeuvred to a new path that, at closest approach, passed 500km from the comet. The flyby spacecraft observed and recorded data about the impact and the ejected material blasted outwards. Meanwhile, on Earth and from orbit above, professional and amateur astronomers using large and small telescopes observed the impact and its aftermath, including changes in the comet's activity.

Analysis of the wealth of data obtained as a result of the Deep Impact mission will take many years, but already scientists are gaining new insights into the structure and composition of the nucleus of comet Tempel-1. The nucleus measured 7.6 x 4.9km and had a surface which displayed both very smooth areas and features that look like impact craters. The solid material blasted out from the impact site was in the form of extremely fine dust and snow, pulverised by the impact. This formed a huge eruption plume which hid the crater produced by the impact from view. It seems that comet Tempel-1 has a very fluffy structure that is weaker than a snowdrift of finely powdered snow. As the hot eruption plume expanded rapidly outwards, scientists looking at the spectrum of the cloud saw emission bands for water vaporised by the heat of the impact. These were followed a few seconds later by absorption bands from ice and dust particles excavated from below the surface and not melted or vaporised. Another interesting finding was the huge increase in carbon-containing molecules detected in spectral analysis of the ejection plume. This indicates that comet Tempel-1 contains a substantial amount of organic material. The nucleus also appears to be extremely porous and this porosity allows the surface of the nucleus to heat up and cool down almost instantly in response to sunlight. This suggests heat is not easily conducted to the interior and the ice and other material deep inside the nucleus may be pristine and unchanged from the early days of the Solar System, just as many scientists had suggested. However, the Deep Impact results suggest that perhaps comets are more like 'snowy dirtballs' rather than the 'dirty snowballs' that Fred Whipple suggested.

Our next close-range look at a comet will be courtesy of the European Space Agency's Rosetta mission. Originally scheduled to rendezvous with periodic comet Wirtanen, the launch of Rosetta had to be delayed due to problems with the launch vehicle, and so a new target had to be found. Eventually launched on 2 March 2004, Rosetta is now heading for periodic comet Churyimov-Gerasimenko and in May 2014 will hopefully go into a 25km high orbit around the cometary nucleus - the first ever comet orbiter. In this way, Rosetta will obtain an extremely detailed map of the entire surface of the nucleus from orbit. In addition to monitoring the activity of the comet as it travels inwards towards the Sun and then back out again, Rosetta will deploy a small lander, called Philae, onto the surface of the nucleus, where it will make a detailed analysis of the composition and structure.

Even with the Deep Impact data, our present knowledge of comets is really only 'skin deep'. The mass, density, strength, internal structure and composition of the nucleus are all still poorly understood. We also do not know whether the physical and chemical properties vary from one location to another on an individual cometary nucleus or from one comet nucleus to another. The four cometary nuclei we have studied so far at close range have all possessed some individual characteristics. Indeed, as yet we have only studied the nuclei of short-period comets. The nucleus of a Great Comet, such as Hale-Bopp which graced our skies in March 1997, may be as much as 35-40km across and we have never

Left: *This spectacular image of comet Tempel-1 was taken 67sec after it obliterated Deep Impact's impactor spacecraft. Scattered light from the collision saturated the camera's detector, creating the bright splash seen here. Linear spokes of light radiate away from the impact site, while reflected sunlight illuminates most of the comet surface. The image reveals topographic features, including ridges and possibly impact craters formed long ago. Image courtesy of NASA/JPL-Caltech/University of Maryland.*

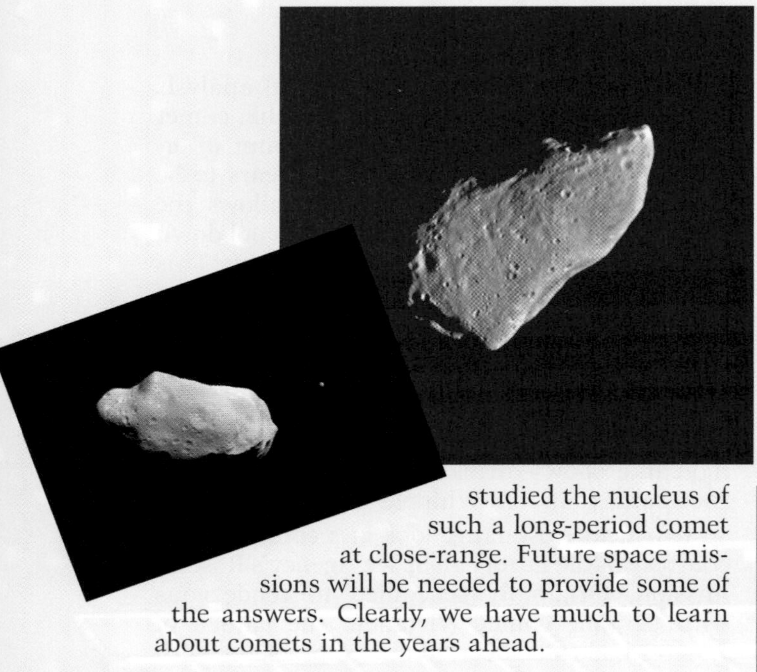

studied the nucleus of such a long-period comet at close-range. Future space missions will be needed to provide some of the answers. Clearly, we have much to learn about comets in the years ahead.

Asteroids: the first close-ups

Our first close-up views of asteroids were mere snapshots obtained by spacecraft while en route to somewhere else. NASA's Galileo mission to Jupiter, due for launch using the Space Shuttle in May 1986, had been planned to include a flyby of the Main Belt asteroid 29 Amphitrite on 6 December 1986 - the first close-range inspection of such a body, and a great scientific bonus for the Galileo project. But it was not to be. The Challenger disaster of January 1986 necessitated a radical redesign of the Galileo mission. This was because the Centaur liquid-fuelled upper stage, which was to send the spacecraft direct to the giant planet and had been adapted for use with the Shuttle, was now deemed too dangerous to be flown to low-Earth orbit in the Shuttle's payload bay. Instead, the much less effective IUS solid-fuelled rocket had to be used.

The revised plan meant that Jupiter could only be reached if the Galileo spacecraft followed a highly circuitous route involving flybys (and associated gravity assists) from Venus once and the Earth twice - the so-called Venus-Earth-Earth Gravity Assist (VEEGA) trajectory. Eventually launched in October 1989, the Venus flyby lifted the aphelion (furthest point from the Sun) of the spacecraft's orbit out beyond Earth's orbit, the first Earth flyby moved it out to the asteroid belt and the second Earth flyby extended it to Jupiter's orbit. Arrival at the giant planet finally took place in December 1995. Now flybys of two Main Belt asteroids were planned while Galileo was en route to Jupiter: 951 Gaspra on 29 October 1991 and 243 Ida on 28 August 1993.

Gaspra is a typical Main Belt asteroid, measuring 20 x 12 x 11km. Its surface reflects about 20% of the sunlight striking it. Gaspra is classified as an S-type asteroid. Such bodies are of a silicaceous (stony) composition and are likely composed of nickel-iron metal mixed with iron- and magnesium silicates. S-type asteroids are most common in the inner part of the Main Belt, but become rarer further out. Gaspra is a member of the Flora family and its irregular, tapered shape is consistent with it being a fragment produced by the break-up of a somewhat larger body, a few hundred kilometres across, sometime in the past.

Ida, another typical Main Belt asteroid, measuring about 56 x 24 x 21km, was more than twice as large as Gaspra. Ida is a heavily-cratered, irregularly-shaped asteroid, again classified as S-type. It is a member of the Koronis family, which scientists believe was created when a larger body perhaps 200 to 300km in diameter was smashed up relatively recently – at least considerably after the Solar System formed some 4.5bn years ago. The big surprise was that Ida turned out to have a tiny moon, now called Dactyl, which measured just 1.2 by 1.4 by 1.6km. It's a 'chip off the old block' – a tiny fragment probably knocked off in an earlier collision, perhaps the one that shattered the larger body which produced Ida itself.

Monoliths or rubble piles?

The first close-range images of Main Belt asteroids indicated that many are probably the result of collisions between larger objects. In such a collision, several large fragments and many much smaller fragments would be produced. When these collisions occur, some fragments may be ejected into orbits which take them closer to the orbit of Mars. In repeated close passages to Mars over many years, fragments may be perturbed such that they enter Mars-crossing orbits. These objects may then approach or even cross the orbit of the Earth, and have the potential to collide with our planet. It is thought that there are probably between 500 and 1,000 asteroids larger than one kilometre across moving in Earth-crossing orbits,

Above: *A mosaic of two images of the asteroid 951 Gaspra acquired by the Galileo spacecraft from a range of 5,300km on 29 October 1991. More than 600 craters, 100-500m in diameter are visible. The number of such small craters compared to larger ones is much greater for Gaspra than for bodies of comparable size such as the satellites of Mars. Image courtesy of NASA Jet Propulsion Laboratory.*

Inset: *A colour picture of the asteroid 243 Ida made from images taken by the Galileo spacecraft from a range of 10,500km on 28 August 1993. The images used are from the sequence in which Ida's tiny moon, Dactyl, was originally discovered. Although appearing just to the right of Ida in this image, Dactyl is actually in the foreground, slightly closer to the spacecraft than Ida is. The satellite is probably about 100km away from the center of Ida. Image courtesy of NASA Jet Propulsion Laboratory.*

Quite often, asteroids are discovered which 'appear' to threaten the Earth at some date in the future, but which after a few days, when their orbits are accurately determined, are found to be perfectly harmless. At present we know of no asteroid which is definitely on a collision course with Earth. So, for the time being, Earth appears safe, but an impact seems inevitable – eventually. Asteroid impacts have clearly happened numerous times since the Earth was formed and it is generally agreed that such events have a dramatic effect on the biosphere. It was time to really get to know the enemy!

Scientists had long wondered whether asteroids were solid rocks (monoliths) or large collections of rubble. Knowledge of the internal composition of asteroids and how well they are bound together would be important if one were ever found to be headed our way, and scientists needed to know how to destroy or deflect it. That was one of the principal reasons behind the Near Earth Asteroid Rendezvous (NEAR) mission to the asteroid 433 Eros – "to knock on them and see what they are made of," to quote the late nuclear physicist Edward Teller. Launched on 17 February 1996, the NEAR mission (later renamed NEAR-Shoemaker in honour of the late planetary geologist Eugene Shoemaker) was the first spacecraft to rendezvous with, go into orbit around, and later land on an asteroid.

Past Mathilde to Eros

The NEAR spacecraft was the first launch in NASA's Discovery Programme. The spacecraft, which was designed and built by The Johns Hopkins University Applied Physics Laboratory, was the shape of an octagonal prism, measuring about 1.7m on a side, with four solar panels and a fixed 1.5m X-band high-gain radio antenna. It carried a suite of instruments consisting of an X-ray/gamma-ray spectrometer, a near-infrared imaging spectrograph, a multispectral camera fitted with a CCD imaging detector, a laser altimeter and a magnetometer. A radio science experiment was also carried out using the NEAR tracking system to estimate the gravity field of Eros.

On its first orbit following launch and prior to its encounter with Eros, NEAR flew within 1,200km of the carbon-rich C-class asteroid 253 Mathilde on 27 June 1997. Mathilde is one of the blackest objects in the Solar System, reflecting only 3% of the light that strikes it. It is blacker than coal, or two times the darkness of a chunk of charcoal. Mathilde turned out to have more large craters pockmarking its surface than either Gaspra or Ida; there were at least five craters larger than 20km across. Mathilde measures roughly 60 x 47km, and is approximately four times the size of Gaspra and two times the size of Ida. Mathilde has an extremely long rotation period of 415hr. Only two other asteroids are known to have longer rotation periods.

Following the Mathilde flyby, NEAR executed its first major deep space manoeuvre which lowered the perihelion of the spacecraft's orbit. NEAR then flew by the Earth on 23 January 1998 for a gravity assist, which increased the spacecraft's orbital inclination and reduced the aphelion distance to match those of Eros, in readiness for the rendezvous and orbit of the asteroid, originally due to take place in January 1999. Unfortunately, a software problem and a temporary loss of communication with the spacecraft caused an abort of the first rendezvous burn on 20 December 1998 and a new mission plan had to be hastily drawn up. Consequently, the spacecraft made a flyby of Eros on 23 December 1998 and, following a series of thruster burns to fine-tune its trajectory and another orbit of the Sun, it finally entered orbit around Eros on 14 February 2000, some 13 months later than planned. Searches for satellites of Eros were carried out prior to arrival, but none were found.

Eros: from orbit to touchdown

433 Eros is a member of the Amor class of Near-Earth Asteroids (NEAs) having an orbit which crosses Mars' path but doesn't intersect that of Earth. It is another S-class asteroid and its dimensions of 33 x 13 x 13km in size make it one of only three NEAs with diameters greater than

Above: *An image mosaic of asteroid 253 Mathilde constructed from four images acquired by the NEAR spacecraft from a range of 2,400km on 27 June 1997. The surface has many large craters, including the deeply shadowed one at the center, which is estimated to be more than 10km deep. The shadowed, wedge-shaped feature at the lower right is another large crater viewed obliquely. The angular shape is believed to be the result of a violent history of impacts. Image courtesy of John Hopkins University Applied Physics Laboratory.*

10km. The images of Eros obtained by the renamed NEAR-Shoemaker craft from a variety of orbits revealed it to be a heavily scarred, vaguely shoe-shaped asteroid. Studies were made of the asteroid's size, shape, mass, magnetic field, composition, and surface and internal structure. Although initially orbiting just over 300km above Eros, this was subsequently reduced to between 100 and 200km, and the craft approached as low as 24km above the surface of the asteroid during the final days of the mission.

The heart of Eros is probably solid, but the surface is clearly a collection of progressively fragmented rubble. Numerous impacts have created at least 100,000 craters on the surface, and some are very large. One puzzle is a lack of small impact craters, so some process must be covering up or eroding them. One idea is that when an impact occurs, shaking caused by the shock waves causes the walls of the smallest craters to crumble and settle, while larger craters are much less affected by this shaking. There are also smooth, flat areas at the bottom of some craters. Boulders are plentiful on Eros. In certain areas, 25 boulders the size of a house or larger sit on every square kilometre. There are estimated to be about a million such rocks in total. Smaller boulders appear even more numerous. Boulders one metre across or bigger are 500 times as plentiful as the house-sized rocks. In general, boulders on Eros appear to be buried in the asteroid's regolith, or soil, to different depths.

All told, the craft returned more than 160,000 detailed pictures of Eros. These images represent by far the most detailed look at an asteroid ever obtained. Its orbital surveillance mission over, NEAR-Shoemaker was brought down to a soft landing on the surface of Eros on 12 February 2001, with the mission ending 16 days later. The spacecraft impacted at a speed of about 5 to 7 km per second, and obtained 69 high-resolution images before touchdown, the final image showing an area 6m across. Although NEAR was not designed as a lander, amazingly it survived the low-velocity, low-gravity impact, and a signal continued after touchdown using the omni-directional low-gain antenna as a beacon. The NEAR-Shoemaker team did not attempt to lift off from the asteroid again and the craft is still there sitting on the surface.

Above: This image mosaic of Eros' southern hemisphere, taken by NEAR-Shoemaker's Multi Spectral Imager on 30 November 2000, offers a long-distance look at the cratered terrain south of where the spacecraft touched down on 12 February 2001. In this view, south is to the top and the landing site itself is just into the shadows, slightly left of centre. Image courtesy of John Hopkins University Applied Physics Laboratory.

Below: NEAR-Shoemaker captured this image of the surface of Eros on 7 January 2001, from an altitude of 35km. The upper half and lower right parts of the image show surfaces with 'typical' rounded craters and large boulders. However, the abruptly edged swath extending from lower left to middle right is remarkably smooth, subdued, and lacking in small-scale detail of any type – as if it had been altered by a giant eraser! The whole scene is about 1.4km across. Image courtesy of John Hopkins University Applied Physics Laboratory.

Enter the falcon

The Japan Aerospace Exploration Agency (JAXA) attempted to go one step further with its Hayabusa asteroid sample-return spacecraft (originally named Muses-C), late in 2005. The ambitious plan called for Hayabusa, which means 'peregrine falcon' in Japanese, to act much like its namesake. First it would hover above its prey – a tiny asteroid called Itokawa – then would 'drop in for the kill', descending to the surface, capturing some samples and returning them to Earth for analysis. In the course of this bold venture, Hayabusa would also test four advanced technologies: an electric propulsion (ion drive) system; an autonomous navigation system; a sample collection system; and a sample return capsule that would re-enter the Earth's atmosphere.

The S-type asteroid 25143 Itokawa was chosen as the target for the mission, because it is a member of the Apollo class of Near-Earth Asteroids (also referred to as Near Earth Objects, or NEOs) whose orbit not only intersects that of the Earth, but is also a 'Mars-crosser'. Itokawa's distance from the Sun varies between 142.6m km and 253.4m km in a period of 556.4 days. It was known to be considerably less than a kilometre across and radar imaging using a radio telescope at the Goldstone Deep Space Communications Complex in California's Mojave Desert had revealed a somewhat elongated shape, in common with many other NEOs. It has a spin period of just over 12hr.

As is the case with many space missions, Hayabusa has had its share of 'nail-biting' moments. The mission had been originally intended to investigate the asteroid 4660 Nereus, with 1989 ML as a back-up. It was also supposed to have carried a NASA Jet Propulsion Laboratory rover, with a sample return landing in Utah, but NASA pulled out late in 2000. Then the planned launch in July 2002 atop a Mu-5 three-stage solid-fuelled rocket had to be postponed, when a similar rocket failed to deliver Japan's Astro-E X-ray observatory to orbit: the rocket veered off course and the satellite burned up in the atmosphere. The resulting delay meant that the original intended targets would be 'out of range' by the time another launch could be scheduled, so a new destination had to be found. Eventually, the asteroid 1998 SF36, discovered on 26 September 1998 by the Lincoln Near Earth Asteroid Research (LINEAR) project[1], funded by the United States Air Force and NASA, was selected. Shortly thereafter, the International Astronomical Union bestowed it with its permanent name – Itokawa. This was highly appropriate because the late Professor Hideo Itokawa is widely acknowledged as the 'father' of the Japanese space program.

Hayabusa was finally launched from the Uchinoura Launch Center at Kagoshima on Kyushu Island on 9 May 2003, but six months later, in November 2003, the spacecraft was buffeted by one of the biggest solar flares in history. It was yet another tense moment for everyone involved with the mission, but Hayabusa emerged unscathed, with the performance of only some solar cells degraded. The following year, on 19 May 2004, Hayabusa successfully made a swingby of Earth, at a distance of just 3,725km, thereby gaining the velocity it needed to reach the asteroid from the received gravity assist. During this Earth flyby, the spacecraft acquired images of the Earth and Moon to test and calibrate the on-board camera called AMICA (Asteroid Multi-band Imaging Camera).

Slow approach and rendezvous

On 31 July 2005, as Hayabusa approached its target, one of the spacecraft's three reaction control wheels (the X-axis wheel) encountered increasing friction as it spun, to the point where it had to be shut down. These reaction control wheels are used to control the orientation of the spacecraft and point the instruments and antennas at their chosen targets. However, Hayabusa was designed to function with only two reaction control wheels working and so was able to resume attitude stability and normal operations immediately. Unfortunately, on 2 October, the Y-axis reaction wheel failed as well and after this failure the orientation of the spacecraft had to be maintained using just the Z-axis reaction wheel and its thrusters, a process that consumed valuable fuel.

By virtue of its ion engines, Hayabusa was able to make a very unusual low-speed rendezvous with the asteroid. From an already slow approach speed of just 38m/sec at a distance of 35,000km, the spacecraft's ion engines were turned off at a distance of 3,500km leaving a residual approach speed of just 10m/sec. This speed was gradually reduced still further by the

The knobbly shape of the tiny asteroid 25143 Itokawa revealed in images from Hayabusa was not what had been expected based on radar studies. Itokawa has hilly and rocky regions as well as some very smooth areas. Such tiny asteroids, with a very low gravitational pull, were not expected to have much surface regolith, so the discovery of dust and rocks on the surface, left over from impacts, was another surprise. Itokawa may well be a rubble-pile asteroid. Images courtesy of Institute of Space and Astronautical Science/ Japan Aerospace Exploration Agency.

onboard reaction control system so that on arrival in the vicinity of the asteroid on 12 September 2005 (and despite being jolted just days before by yet another intense solar flare), Hayabusa came to a standstill with respect to Itokawa, on a line roughly between the asteroid and the Sun, initially at a distance of 20km. Later this distance was reduced to just 7km. While 'hovering' above the surface, Hayabusa's instruments provided estimates of the asteroid's mass and bulk density, studied the surface in great detail and determined which minerals are present.

Images of Itokawa acquired by Hayabusa both during the approach phase and while hovering above the asteroid confirmed earlier findings. With dimensions of just 540 x 270 x 210m, Itokawa is, by far, the smallest asteroid surveyed to date, and it is indeed about twice as long as it is wide, with a distinctive 'waist' around the middle. Such observations suggest that Itokawa may be a contact binary, formed by two (or more) small asteroids that have gravitated towards one another and have stuck together. Indeed, a smooth area, tentatively dubbed the 'Muses Sea' by mission scientists, may mark the area of contact. This shape and structure is typical of many other NEOs that have been imaged using radar during close approaches to the Earth, such as the dumbbell-shaped objects 4179 Toutatis and 4769 Castalia. Itokawa's mass is estimated to be just 4×10^{10}kg, and its density about 2.3gm/cm^3. The Hayabusa images also revealed a surprising lack of impact craters and a very rough surface covered with boulders many metres in diameter, indicating that Itokawa is most likely a 'rubble pile' formed from fragments that have come together over time.

One problem after another

The plan was for the Hayabusa spacecraft to collect up to three surface samples of the asteroid using a sample collection horn that would catch small pieces of the asteroid sprayed out when tantalum pellets were fired into its surface at 300m/sec. These surface samples would then be stowed in the spacecraft's sample capsule for subsequent return to Earth. The initial descent towards the surface of the asteroid went well, but due to problems with the optical navigation system the 4 November 'rehearsal' landing on Itokawa was abandoned and rescheduled.

The plan also called for a tiny surface hopper, called MINERVA, to take advantage of Itokawa's extremely low surface gravity by hopping considerable distances across the surface of the asteroid acquiring surface temperature measurements and high-resolution images with its miniature cameras. Unfortunately, when the lander was released on 12 November 2005, the probe was ascending and at a much higher altitude than intended and instead of dropping onto the asteroid's surface MINERVA escaped Itokawa's gravitational pull and tumbled into space.

On 19 November, Hayabusa finally touched down on the surface of Itokawa, but there was considerable confusion about exactly what had happened both during and after the manoeuvre, mainly caused by communications problems with the spacecraft. Only on 23 November, after regaining control and communication with the probe and after data from the landing attempt were downloaded and analyzed, was it confirmed that the probe had indeed landed on the asteroid's surface. Unfortunately, the sampling sequence was not triggered since the probe tried to abort the landing, but there is a chance that some dust may have been whirled up into the sampling horn when the craft touched down on the asteroid, so the sample canister attached to the sampling horn was sealed.

On 25 November, a second landing took place, and this time the sampling sequence was triggered. Unfortunately, due to a leak in a thruster the craft was put into a safe mode soon afterwards and control and communiaction were not regained until some days later. Continuing problems with the spacecraft's thrusters led to further breaks in communications, but each time JAXA controllers managed to regain attitude control and re-establish communications. However, when telemetry was obtained and analyzed, it was found there is a strong possibility that the sampling sequence during the landing on 25 November may not have gone as planned because the pellet used to kick up samples from the surface apparently did not fire.

Finally, on 8 December 2005, a fuel leak caused a sudden change in the craft's attitude, leading to a total loss of communications. It would take a month or two for the spacecraft to stabilize, and then the rotation axis must be directed towards the Sun and Earth within a specific angular range for communications to be re-established. The probability of this is 60% by December 2006, and 70% by the spring of 2007.

In the meantime, the craft has missed the opportunity of returning to Earth as originally planned. Hayabusa, still hovering near Itokawa, needed to fire its main engine by mid-December 2005 in order to enter a trajectory that would bring it back to Earth in June 2007. However, the continuing thruster problems forced controllers to reschedule the spacecraft's return. Assuming that control of the craft can be regained, the new plan would see Hayabusa return to Earth, with its sample return capsule, in June 2010, three years later than originally intended. Unfortunately, there is no guarantee that the spacecraft has actually collected any samples from the asteroid's surface!

Whatever the contents of the sample return module when it finally parachutes down in the Australian Outback in 2010, Hayabusa has already been an outstanding success. It has played a key role in advancing our knowledge of NEOs by showing that there are important differences in the composition and surface geology of small asteroids compared with larger ones. Clearly further detailed investigations of NEOs will be essential to improve our understanding of these diverse and fascinating objects.

John Mason *is Subject Advisory Editor for Praxis Publishing's books in Space Exploration, and he is a leading British writer, lecturer and broadcaster on astronomy and spaceflight.*

(Footnotes) [1] The LINEAR project uses a pair of one-metre diameter telescopes at Lincoln Laboratory's Experimental Test Site on the White Sands Missile Range in Socorro, New Mexico. The goal of the LINEAR program is detect and catalogue Near-Earth Asteroids that threaten the Earth.

Artist's impression of ESA's Rosetta spacecraft in orbit around the nucleus of comet Churyimov-Gerasimenko. Also shown is the lander Philae, shortly after deployment from the orbiter, descending towards the surface of the nucleus through the dust and gas of the surrounding coma. Image courtesy European Space Agency.

5.1 A vision for SPACE EXPLORATION...

This chapter reviews the biggest plan for space exploration proposed since the time of Apollo - the Vision for Space Exploration announced by President Bush. In the first part of this chapter, *John Catchpole* outlines the origins, key features and critical paths of the new plan.

Rolling out the BUSH PLAN

WHEN *COLLIERS Magazine* first published it in America, in the early 1950s, it must have seemed incredible, nothing short of science fiction. Only the author and similar far-sighted visionaries recognised it for what it was – a blueprint for human exploration of the Solar System.

It is a bold plan, in four stages:

- ***Develop a reusable crewed spacecraft***
- ***Construct and permanently occupy a space station in Earth orbit***
- ***Establish a base on the Moon to support a permanent human presence***
- ***Send humans to Mars and return them safely to Earth***

Bold as it was, the plan was derailed by the Cold War. In the race to put the first human into orbit, the re-usable spacecraft was replaced by the one-shot ballistic capsule launched on a one-shot ballistic missile. Humans landed on the Moon in 1969, but the Project Apollo was a politically, rather than scientifically driven. Some critics like to refer to one of humankind's greatest achievements as nothing more that "footprints and flags." Russia lost both the "space race" and the Cold War. A decade after Soviet Communism imploded, the Russian economy is still in ruins and her space programme is annually left desperately underfunded by a government struggling to do best by its people.

Having placed twelve men on the Moon, America abandoned Project Apollo once it had served its political purpose. No humans have walked on the Moon since 1972. America then developed the re-usable Space Shuttle. It bears remarkable similarities to the re-usable spacecraft in the 1950s *Colliers Magazine* articles. Having also won the Cold War, America invited her old advisories and several of her long-term allies to join her in constructing the International Space Station (ISS), which has been permanently occupied in Earth orbit since November 2000.

Origins

With the Cold War over, the 1950's plan for human exploration of the Solar System seems to have come back on track, but all is not well. America has a re-usable crewed spacecraft, the Space Shuttle, but fourteen astronauts have been lost in two very public disasters. The ISS is permanently occupied, but it only goes around in circles. It attracts very little political support and even less public interest.

To get ISS crews back in an emergency, NASA tried to develop X-38 prototype Crew Return Vehicle (CRV), but it was cancelled by President George W. Bush in 2001, in an attempt to bring the ISS budget under control. Next was the Orbital Space Plane (OSP). The OSP was to have served as both a Crew Transfer Vehicle and an emergency CRV. Two groups of American aerospace companies began developing their designs for the OSP. Both released artwork showing a re-usable lifting-body concept for their OSP, but, interestingly, one also completed a study of a ballistic capsule design. At the same time, a group of ex-Project Apollo managers and engineers completed a study of a new spacecraft employing the outward design of the old Apollo Command and Service Module, thereby allowing the original Apollo engineering data to be used in its development. This was only a paper study and not a serious attempt to design an OSP. Throughout its short life OSP was heavily criticised for being too narrowly focused on the ISS role and not offering any potential for future development.

Following the loss of the second Space Shuttle, Columbia, on 1 February 2003, some people demanded that it was time to stop flying around in circles and to re-capture a spirit of exploration. Supporters of a strong space programme called for humans to accept the extraordinary challenge that is the next step - a human expedition to Mars. At the other extreme of the same debate, there were also demands for the human space programme to be abandoned and the money spent on space exploration redirected, preferably to social programmes inside America.

The announcement

President George W. Bush addressed the nation from NASA Headquarters in Washington D.C., on 14 January, 2004. The speech that he made was the culmination of many months of work by White House staff, NASA and numerous other agencies to give NASA new goals for the future.

"Today I announce a new plan to explore space and extend a human presence across our Solar System. We will begin the effort quickly, using existing programmes and personnel. We'll make steady progress, one mission, one voyage, one landing at a time...

"Our first goal is to complete the International Space Station by 2010. We will finish what we have started. We will meet our obligations to our 15 international partners on this project. We will focus our future research aboard this station on the long-term effects of space travel on human biology. The environment of space is hostile to human beings... Research onboard the station and here on Earth will help us better understand and overcome the obstacles that limit exploration.

"To meet this goal, we will return the Space Shuttle to flight as soon as possible...The Shuttle's chief purpose... will be to help finish assembly of the International space Station. In 2010, the Space Shuttle, after nearly 30 years of duty, will be retired from service..."

It was established that Columbia had been lost due to a piece of foam detaching itself from the External Tank (ET) during lift-off and striking the leading edge of the left wing. The resulting hole was not visible from the crew compartment windows and the crew completed their fourteen day mission unaware that it was there. During re-entry, superheated plasma entered the hole in the wing causing it to fail and ultimately tear away. Columbia and her crew were doomed.

The Space Shuttle returned to flight briefly in 2005, but was grounded immediately, after a single flight. Cameras had shown that the ET was still shedding foam. The Space Shuttle will be returned to flight. NASA's astronauts will be prepared to take the risks involved in flying the Space Shuttle in order to complete the construction of the ISS. When that is done the venerable Space Shuttle will be retired and it will be many decades before we see its like again. President Bush continued:

"Our second goal is to develop and test a new spacecraft, the Crew Exploration Vehicle, by 2008 and to conduct the first manned mission no later than 2014. The Crew Exploration Vehicle will be capable of ferrying astronauts and scientists to the space station after the Shuttle is retired. But the main purpose of this spacecraft will be to carry astronauts beyond our orbit to other worlds..."

NASA Administrator Sean O'Keefe was sure that much in the OSP studies would be relevant to the new Crew Exploration Vehicle (CEV) so he asked the two aerospace contractor groups to complete their OSP studies before undergoing any realignment to commence development of the CEV. In O'Keefe's mind both groups would develop and build there own CEV, which would be launched on an Extended Expendable Launch Vehicle (EELV), either an Atlas-V or a Delta-IV. Each vehicle would then undergo a fly-off, a series of flight-tests to see which one NASA would choose to develop as the operational CEV. It was a practice common in the procurement of high-performance aircraft for the American military.

In time, Lockheed Martin released details of a re-usable winged CEV, with a separate pressurised Mission Module and propulsion stage for lunar flights. Boeing released no details of its CEV, but neither did it remove the artwork of the capsule-based OSP from the company website.

Griffin's vision

Sean O'Keefe retired as NASA Administrator in February 2005. His position was filled by Michael Griffin, a former senior NASA engineer, on 14 April 2005. Griffin came with a reputation. He was a vocal critic of the Space Shuttle and the ISS. He believes that NASA had been stuck in Low Earth Orbit for too long and that the Administration's role is to explore the Solar System and the Universe as a whole. Griffin probably would not want to admit it, but his attitude is summed-up by the Mission Statement issued by Sean O'Keefe at the start of his Administration. It read:

The new NASA mission for the future is:

- ***To improve life here.***
- ***To extend life to there.***
- ***To find life beyond.***

The NASA mission is:

- ***To understand and protect our home planet.***
- ***To explore the Universe and search for life.***
- ***To inspire the next generation of explorers – as only NASA can.***

Griffin also believed that NASA should make the most of existing Space Shuttle technology in developing the next generation of launch vehicles. Doing so would reduce the cost and time involved in their development because the Space Shuttle hardware is already rated to carry humans into space. In September 2005, Griffin's NASA released details of its anticipated new hardware.

The Crew Exploration Vehicle

A capsule-based Crew Exploration Vehicle (CEV) will be developed first, along with a new Crew Launch Vehicle. The CEV will consist of a conical Crew Module (CM) and cylindrical Service Module (SM). It will superficially resemble the Apollo Command-Service Module, the portion of the Apollo spacecraft that remained in lunar orbit. The CEV will contain everything required to support four astronauts on a lunar voyage and six astronauts on a later journey to Mars. In the first instance the CEV will be used to ferry crews to and from the ISS. It will also serve as an American emergency Crew Return Vehicle. An un-crewed cargo CEV will also be developed allowing the new spacecraft to serve a similar role to Russia's Soyuz / Progress spacecraft. At launch, an Apollo-style tractor rocket Launch Escape System will be mounted on the Apex of the CM, to pull it and its human cargo away from a malfunctioning launch vehicle.

Like Apollo before it, the new CM will be a cone with the crew access hatch and windows in its side. The curved ablative heatshield on the wide end will be 5.5m in diameter, giving the CM three times the volume of the Apollo Command Module and a mass of 25 tonnes. The docking system will be mounted at the apex and, like in Apollo, an internal transfer tunnel will be surrounded by the recovery parachutes. Manoeuvring thrusters will be mounted around the base and apex. Following re-entry, three parachutes will be deployed and the heatshield will be jettisoned to allow air-bags to deploy and absorb the final landing loads. Unlike Apollo, the CM will touch down on land in north America. Splashdown will be a back-up to land-landing. The heatshield will be replaced after each flight.

The SM will contain a liquid oxygen / liquid methane Service Module Propulsion System, which has yet to be developed. It will also hold the tanks for the oxygen, water and propellant that will be consumed during the flight. Manoeuvring thrusters will be mounted in four quads on the outside of the SM. Electrical power for the CEV will come from two photovoltaic arrays that will be folded for launch and deployed after the spacecraft separates from its launch vehicle. As with Apollo, the CM and SM will serve as one spacecraft until just prior to re-entry. The Service Module Propulsion System will be used for the de-orbit burn, after which the SM will be jettisoned to burn-up in the atmosphere.

Above: *Return to the Moon – outbound. (Top) Crew Launch vehicle taking crew to Earth orbit. (Middle) Cargo Launch vehicle with lunar lander and Earth departure stage. (Bottom) Crew vehicle docked with lander and departure stage leaving Earth orbit, bound for the Moon. Artist's concepts courtesy NASA/John Frassanito and Associates.*

The first stage of the Crew Launch Vehicle (CLV) will be a single four-segment Space Shuttle Solid Rocket Booster. A new second stage will be developed, with a single Space Shuttle Main Engine (SSME), burning liquid hydrogen and liquid oxygen. This CLV will lift 25 tonnes into Low Earth Orbit. A fifth segment in the SRB will add an extra seven tonnes to the lift capability. At launch, the Solid Rocket Booster will fire until its fuel is expended. It will then be jettisoned. If not required, the LES will be jettisoned at this time. The liquid propellant second stage would then fire to place the CEV into Earth orbit, where it will separate and deploy its antennae and photovoltaic arrays.

NASA estimated that, due to the in-line design, with the CEV placed at the top of the stack, thereby allowing the use of a tractor rocket LES, the new CLV / CEV will be 10 times safer than the present Space Shuttle. The new CLV / CEV combination is planned to make its first automated flight in 2008, with its first crewed flight in 2011, just one year after the Space Shuttle is due to be retired. This is three years earlier than demanded by the President's speech and by Sean O'Keefe's plans.

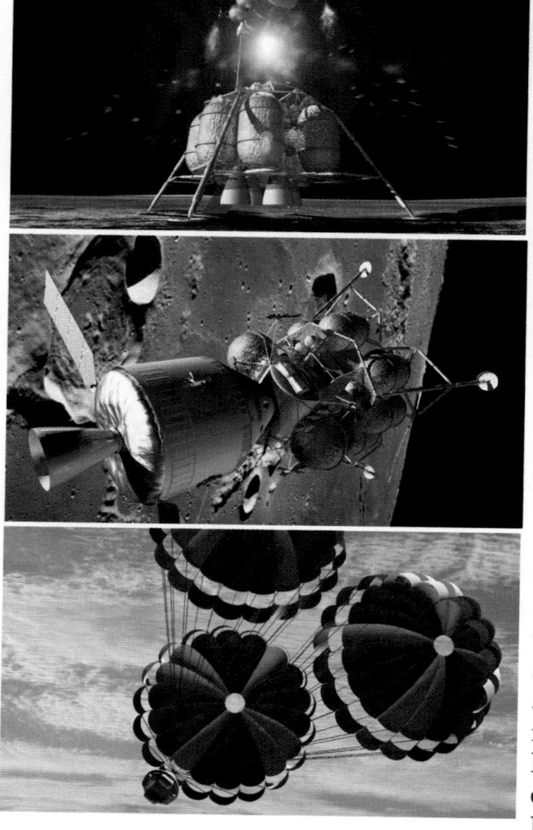

Lunar lander

The President went further:

"Our third goal is a return to the Moon by 2020, as the launching point for missions beyond. Beginning no later than 2008, we will send a series of robotic missions to the lunar surface to research and prepare for future human exploration. Using the Crew Exploration Vehicle, we will undertake extended human missions to the Moon as early as 2015, with the goal of living and working there for increasingly extended time...The Moon is a logical step toward further progress and achievement. With the experience and knowledge gained on the Moon, we will then be ready to take the next steps of space exploration: human missions to Mars and worlds beyond.

"A human return to the Moon will be preceded by robotic flights, which will help select landing sites and determine what natural resources are available on the lunar surface. These flights will begin in 2008 and continue through 2011".

The reason behind NASA's September 2005 release was to show how humans would be returned to the Moon, in 2018, two years earlier than in the President's speech. The new plan requires two new launch vehicles and two new spacecraft, including the CLV and CEV. The second two vehicles are the Lunar Heavy Cargo Launch Vehicle (LHCLV) and the Lunar Landing Vehicle (LLV). Definition of these two vehicles will begin in 2011, with work beginning in 2013 on the equipment required to sustain long-duration stays on the lunar surface.

NASA artwork shows the LLV as an open framework, four legged vehicle with a descent engine surrounded by insulated propellant tanks and equipment storage. The descent engine will burn liquid hydrogen and liquid oxygen. The Ascent Stage will contain the pressurized Crew Compartment and airlock for access to the lunar surface, as well as its own rocket motor and propellant tanks. That rocket motor will burn liquid methane and liquid oxygen. As with the Apollo Lunar Module, the Descent Stage will act as a launch pad. Unlike the other vehicles, the September 2005 announcement gave very few details of the LLV.

Above: *Return to the Moon – there and back. (Top) Crew vehicle and lander in lunar orbit. (Middle) Astronauts blast off from the lunar surface in the ascent stage. (Bottom) Parachutes deployed after re-entry. Artist's concepts courtesy NASA/John Frassanito and Associates.*

Heavy lifter

In the LHCLV, a pair of five-segment Space Shuttle Solid Rocket Boosters (SRB) will be mounted on either side of a new liquid propellant Central Core based on modified Space Shuttle External

Tank (ET) technology. Five Space Shuttle Main Engines, at the base of the central core, will burn liquid hydrogen in liquid oxygen. The payload will be placed within a launch shroud at the top of the core. This vehicle will be capable of launching 160 tonnes into low Earth orbit.

The second stage, called the Earth Departure Stage (EDS), will have two J-2S rocket motors which will burn liquid hydrogen and liquid oxygen. These will be derived from the re-startable J-2 motor used on the upper stages of Apollo's

Saturn-V. On the new lunar journey, the EDS will act as a second stage, delivering itself and the LHCLV's cargo into low Earth orbit. On the lunar mission, the payload will be the un-crewed LLV. In this configuration, the LHCLV will be able to place 125 tonnes into Earth orbit.

Lunar mission

At launch, the LHCLV's two SRBs and the Central Core will ignite together. When their propellant is depleted, the SRBs will be jettisoned while the Central Core continues the climb. The payload shroud will be jettisoned. Following the depletion of its liquid propellant, the Central Core will also shut down and be jettisoned. The EDS will ignite and complete the flight into orbit. The EDS will shutdown, having depleted only part of its propellant supply.

With the EDS and LLV in orbit, the CLV / CEV combination will be launched. Having separated from its launch vehicle, the CEV will perform a rendezvous and dock with the LLV, mounted in the end of the EDS. Following systems checks, the two J-2S rocket motors on the EDS will be re-started to push the spacecraft out of Earth orbit. The CEV/LLV will separate from the EDS, which will play no further role in the flight. The forces of nature will carry the spacecraft towards the Moon. Mid-course corrections will be made using the SM's thrusters. The Service Module Propulsion System will slow the spacecraft down and let gravity capture it into lunar orbit.

Unlike Apollo, when one astronaut remained in lunar orbit, all four crew members will enter the LLV and perform preparations for the landing. The CEV, containing everything required for the

return to Earth, will remain in orbit. When the LLV is undocked, it descends to the lunar surface. Landing will be computer controlled with the ability for manual override in the last few hundred metres. Surface activities will consist of performing experiments and geological traverses in anticipation of the day when humans might fly to Mars. At the end of the surface exploration, the Ascent Stage will carry the crew back into lunar orbit where it will carry out a rendezvous and docking with the CEV. Following crew transfer, the Ascent Stage will be jettisoned. The Service Module Propulsion System will perform the burn to take the CEV out of lunar orbit for return to Earth.

As it approaches Earth, the SM will be jettisoned and the CM will turn so that the ablative heatshield covering its broad end faces forward. Following re-entry, landing will be in north America, with the CM descending beneath parachutes and deploying airbags to absorb the final landing forces.

The earliest Moon flights will support a four to seven-day stay on the lunar surface. Later flights will build on the experience of those gone before. Experience on Russia's Salyut and Mir

■ **Above left:** *NASA's new Cargo Launch (far right) and Crew Launch (second from right) rockets compared with the Shuttle and a Saturn V. Artist's concepts courtesy NASA.*

■ **Above:** *Impression of NASA's new Crew Exploration Vehicle (CEV) docked with the lunar lander. The CEV service module is identical in shape (but not in size) to its Apollo predecessor, but unlike the Apollo SM, the new CEV SM, which will be shorter in height and will feature a pair of deployable Soyuz-like solar panels, eliminating the need to carry fuel cells. Artist's concept NASA/John Frassanito and Associates.*

space stations and the ISS will allow a permanent human presence on the Moon. Ultimately, astronauts can expect to spend up to six months on the lunar surface in a permanent structure and will rotate in and out of the duty team just as the Space Station crews do – the new crew will fly a spacecraft to the Moon and the old crew will fly it home.

Inspiring people

This plan has been criticised in the media as being too like Apollo, but in England we have a saying, "If it isn't broken, don't fix it." The Apollo mission profile worked well. The Space Exploration Initiative, as "The Bush Plan" is officially called is about more than spacecraft architecture and returning to the Moon. It is about inspiring people.

As part of his speech, President Bush established the Presidential Commission on Implementation of the United States Space Exploration Policy. Between them, the nine members of the Commission, under Chairman Edward Aldridge Jnr, had relevant and comprehensive experience in space operations, space science, government service, the military, industry and academia. The Commission took evidence at a series of five televised public hearings and established a website that received over 6m hits and over 6,000 formal written submissions. Public comments in support of the new space exploration vision outweighed those against it by a ratio of 7 to 1. The final report was sent to the White house with a covering letter dated 4 June, 2004. It was entitled, *"A Journey to Inspire, Innovate and Discover."* The Commission began by asking, *"Why Go?"*

The Commission answered with three reasons:

- *Exploration: The impulse to explore the unknown is a human imperative… This endeavour presents an opportunity to inspire a new generation of American explorers, scientists, entrepreneurs and innovators who will provide American leadership to the world…Despite the spiritual, emotional and intellectual appeal of a journey to space – exploration and discovery will perhaps not be sufficient drivers to sustain what will be a long and at times risky, journey. We must also undertake this mission for pragmatic, but no less compelling reasons, which have everything to do with life here on Earth.*
- *Growth: Further space exploration will generate new jobs within current industries and will likely spawn entire new markets involving leading-edge manufacturing and flight-support services. The vision requires a large, high-tech manufacturing base and a skilled workforce… The public record is rich with data about how aerospace technology and pure space research routinely spur other tangible advances and unrelated economic opportunity.*

■ ***Above:*** *Impression of a future Moon landing with astronauts working on the lunar surface. By going to the Moon for extended periods of time, astronauts will search for resources and learn how to work safely in the harsh environment. The Moon also offers many clues about the time when the planets were formed. Artist's concept NASA/John Frassanito and Associates.*

■ *Security: Implementation of the vision is also important to the nation's domestic, international and economic security...Much of the United States' current military strength and economic security rests on our technology leadership. Our technological and industrial base must be constantly renewed.*

The Commission suggested: "The Space Exploration Vision can be a catalyst for a much-needed renaissance in math and science education in the United states...Going to the Moon, Mars and beyond may be our destination and our goal, but the journey towards this goal and what it means here on Earth, is what really matters. Space exploration is an opportunity to invest meaningfully in America. It is a much-need opportunity to revitalize our industrial base and nurture the skills needed to drive a new generation of American innovation."

The report suggests that the Space Exploration Initiative must be driven by three imperatives if it is to be successful. The commission suggests it must have:

■ *Sustainability: over several decades, through several Presidential Administrations, numerous Congresses and at least two-generations of American taxpayers. The report says that, "public ownership of this agenda must be broad, deep and non-partisan."*

■ *Affordability: through a progressively complex series of missions, each building on those that had gone before, the Exploration Initiative will progress through "measurable milestones" on a "go as you can pay" basis. The Commission also suggested the use of "significant monetary prizes tied directly to the vision" to drive forward private enterprise and accelerate the development of aerospace technologies required for the journeys to the Moon and Mars. They also supported international involvement in support of the Exploration Vision.*

■ *Credibility: by NASA accepting full and transparent stewardship for the Space Exploration Vision and ensuring that their actions and the vision itself are worthy of the support of Congress and the public. NASA must be prepared to "embrace change with a passion...Budgets must be honoured, schedules and performance milestones successfully met and safety treasured... NASA and its partners must implement robust safety precautions. But NASA, its partners and the public must also acknowledge that our bold resolve to return to the Moon and travel to Mars cannot be drained of all risk."*

The Commission states:

Congressional support for the space exploration vision is crucial. Without Congressional consensus the vision will be stillborn.

We must not underestimate the difficulty and the risk of achieving the goals and benefits of the space exploration vision. It is difficult to hold the attention of the public in today's data-rich and rapidly changing environment, but we need their commitment. There must also be a commitment between Administrations and Congress for adequate and sustained long-term funding. We must have these commitments if we are to succeed in achieving the exploration objectives in the decades ahead.

We must accept failure and risk along the way – because we face unknown environments and we are pushing the state of the art in technology – not because of preventable mistakes or incompetence. We must transform the operation of government agencies, requiring difficult changes in culture, organizational structure and cooperation. The government must be prepared to shed programmes that do not contribute to the fulfilment of the vision.

We must challenge and rely heavily upon the private sector – major companies, small business and entrepreneurs – beyond anything that has ever been attempted in a major government-run programme. The government must execute only those activities that are too risky for private sector involvement. The government must change its focus to provide incentives for the commercialization of space and to create, nurture and sustain a robust space-based industry.

We must think about our country and its future in competitive terms. The future is for our children and they must be trained to sustain this nation's quality of life in a more competitive world, through technological achievement and economic growth. We must reverse the decline in students entering into technical fields and the shortage of well-trained science teachers. We must take advantage of the unique opportunity afforded by this vision to inspire our youth and our teachers to focus on mathematics, science and engineering education.

We must ask and answer bold questions about our origins and our future. We must ponder and innovate and search the depths of space to know our place in the cosmos.

These were the very same reasons that President Kennedy and President Johnson had used to justify Project Apollo in the 1960s.

Mars?

But what about a human flight to Mars? There was no timetable in President Bush's speech for a human flight to Mars and there were no specific details on a human flight to Mars in NASA's September 2005 announcement. NASA will use the same spacecraft architecture to send six astronauts to Mars in the CEV and a re-named Mars Landing Vehicle (MLV). The MLV will produce its own methane rocket fuel for the Ascent Propulsion System from the Martian atmosphere. What the President said was:

This is only the beginning. Funding decisions will be guided by the progress that we make in achieving these goals.

George W. Bush knew that he will not be President when the hard decisions about Mars have to be made. He has left the human missions to Mars in the hands of his long-term successors and he has given them a get-out clause. His successors are free to say, *"We cannot afford it at this time"*, thereby delaying sending humans to Mars until they too have left office and the decision becomes someone else's problem. This is exactly why American astronauts have not walked on the Moon for the past forty four years - no-one in a position of power would make the decision to spend the money to get out of low Earth orbit.

Before a human walks on Mars there will, no-doubt, be new natural disasters and new wars and they will be here, on Earth. In such circumstances a human walking on the surface of Mars becomes of little importance to most people, those people who do not prepare, or fly rockets and spacecraft for a living. It will only take one President with no sympathetic leaning towards the human space programme to cancel the humans on Mars programme indefinitely. All they will have to say is, *"We cannot afford it. Therefore, we will not go."*

Whose plan?

One day, sometime in an as yet undefined future, humans will walk on the surface of Mars. When they do, "The Bush Plan," will have been completed, because up until now no-one has defined the meaning of, *"worlds beyond."* Of course, President George W. Bush only gave voice to the plan. He must have endorsed its content, but he did not write it. The words were written, argued over and edited by White House speech writers, taking their lead from NASA officials who advised them what it is possible to achieve.

But, which President Bush does the plan belong to? On 20 July, 1989 President George Bush senior promised, "First...Space Station Freedom, our critical next step in all our space endeavours. And next, for a new century, back to the Moon, back to the future; and this time, back to stay. And then a journey into tomorrow, a journey to another planet – a manned mission to Mars."

Sound familiar?

In 1989 the plan progressed no further than renaming Space Station Freedom the International Space Station. Congress refused to vote the necessary funding to return human astronauts to the Moon and to let them progress on to Mars. What was this doomed plan called? You guessed it - the Space Exploration Initiative. So perhaps "The Bush Plan" should be called, "The Bush Family Plan."

Alternatively, it could be called, "The Wernher von Braun Plan", for Wernher von Braun wrote the original blueprint for the human exploration of the Solar System. It was a bold plan, in four stages:

- ***Develop a reusable crewed spacecraft.***
- ***Construct and permanently occupy a space station in Earth orbit.***
- ***Establish a base on the Moon to support a permanent human presence.***
- ***Send humans to Mars and return them safely to Earth.***

When *Colliers Magazine* first published it in America, in the early 1950s, it must have seemed incredible, nothing short of science fiction. Only the author and similar far-sighted visionaries recognised it for what it was.

On 14 January 2004, President George W. Bush Junior instructed NASA to turn von Braun's plan into science fact. He also added some words of inspiration:

The vision I outline today is a journey, not a race. I call on other nations to join us on this journey, in the spirit of cooperation and friendship.

So let us continue the journey.

John Catchpole *is a writer on spaceflight and regular contributor to the magazine Spaceflight.*

lander and departure stage prepare to leave Earth orbit, bound for the Moon. Artist's concept courtesy NASA/John Frassanito and Associates.

5.2

...Continuing a vision for SPACE EXPLORATION

In the second part of this chapter, space writer *Jim Oberg* takes a critical look at the Vision for Space Exploration and asks: is there a better way of doing this?

Stepping stones to MARS

a new strategy

FOR MORE than thirty years, the only human space voyages have been going round and round, in the region of space called Low Earth Orbit – perhaps 500km above Earth's surface. Voyages beyond this zone, back out to the moon and ultimately farther, will require surmounting both technological and conceptual barriers.

Yet however hard it was to send astronauts to the Moon, that world now looks practically in Earth's back yard, in Solar System terms. Mars is so much farther away, needing so much faster and long-lived spacecraft and so much more persistent political commitment, that the very idea of a human Mars expedition (and of its budget) has grown more intimidating year by year.

Early in 2004, the White House released a *Vision for Space Exploration* a long range commitment to venture again beyond LEO, out towards the Moon and beyond. It was an inspiring vision, but how it would be accomplished was not yet in focus. And in the following months, the shift in leadership at NASA left policy decisions drifting aimlessly. By coincidence, a new strategic approach to space exploration – a piece-by-piece "bite size" plan that incorporates and expands on similar earlier suggestions – had been under development and has now appeared in public. The engineering and cultural challenges of sending astronauts to Mars are broken into segments that still provide significant scientific and political payoffs. A consensus is forming – this approach might just be 'do-able'

There are useful historical analogies. Only on his third voyage to America did Columbus actually reach the mainland, replicating a pattern of voyaging set by the first Norse explorers of Vinland and precursing the pattern of later European voyages to Antarctica. And all these voyages had been preceded by smaller expeditions to islands just off the European coast.

For other worlds, in a sample size of one, the same pattern holds. Only on the third manned flight to the moon and after two dozen sorties into low earth orbit, did Apollo astronauts actually attempt the landing. Reasoning both by analogy and by unique Martian logic, for decades some space planners have recommended that the next human expeditions beyond low earth orbit scout out

Above: *Artist's concept of a future landing on the Moon. Image courtesy JAXA.*

Left: *Artist's impression of humans on Mars and some of the hardware which might be utilized in the setting up of a Mars base. In the background, Pavonis Mons, a large shield volcano on Mars' equator overlooks the ancient water eroded canyon in which the base is located. Artist's concept courtesy Pat Rawlings and NASA/JPL.*

the 'offshore waters', then nearby islands and only then aim for Mars. And the first spaceships sent there would not land on the planet itself, but rather go into orbit around it and visit its two small moons Phobos and Deimos.

Step by step

Now a new strategy has emerged that lays a concrete theoretical foundation for such a step-by-step approach. The arguments are no longer by analogy, or from aesthetics and instinct – they rest on solid reality. Wesley Huntress, emeritus space scientist for NASA, President of the 'Planetary Society' and now director of Carnegie's Geophysical Laboratory in Washington, DC, originally disclosed details of the new study in his testimony before the Senate committee overseeing space, in October 2003. "America has the right stuff, but today's human space flight program isn't giving the public what it wants," he began. "The whole point of leaving home is to go somewhere, not to endlessly circle the block."

Huntress described a private effort he was then involved in, under the auspices of the International Academy of Astronautics, called "The Next Steps in Exploring Deep Space". Its purpose, he explained, "is to provide a logical and systematic roadmap for the long-term scientific exploration of the solar system beyond low Earth orbit with a goal to land humans on Mars sometime in the next 50 years." The study, he continued, "envisions the establishment of a permanent human presence in space using an evolutionary approach to the development of space transportation infrastructure utilizing well-defined intermediate destinations as stepping-stones to Mars."

Space history already shows us two extremes of 'space strategies'. At one end is Apollo, a magnificent science mission whose infrastructure collapsed immediately after it had provided brief access to the lunar surface. At the other end is the International Space Station, which has offered magnificent promises that have been delayed year after year by the need to assemble immense infrastructure prior to beginning significant science returns. In the most common

conceptions, a human mission to Mars would be "the worst of both worlds" -- an "ISS-class" infrastructure in the extreme, with many, many years (stretching over several different presidential administrations) of hardware development and testing prior to the return of the first science and ending soon afterwards.

A set of destinations

However, this new study describes a more attractive approach. It identifies several specific "destinations" beyond LEO but short of the ultimate Mars landing. First is a 'gateway' zone at the Sun-Earth Lagrangian Point 'L2' (about 1.6m km 'down sun' from Earth). Then are sorties to one or more of the small asteroids known as 'near-Earth objects'. Next are visits to the two moons of Mars, Phobos and Deimos. Reaching the surface of Mars is at the limit of its vision. "There is no single destination for human exploration, as was the case during the Apollo era," the report explains. "There is a set of destinations that is scientifically and culturally compelling."

In this view, such an approach is not only technically superior to an all-or-nothing Mars landing mission, but is more practical in political terms. "We assert that a 'brute force' approach that would jump directly to Mars from our current limited human capability in low-Earth orbit is untenable," the report stated, "and that the annual investment and mission risk required for such a leap are simply too great to be tolerable in today's environment."

Instead, the report calls for "a logical series of steps that will lead humankind progressively deeper into the solar system and ultimately to Mars, with significant scientific discoveries possible at every destination along the way." Each step, taking a number of years, will bring its own rewards: "[I]n the process, they provide important opportunities for scientific discovery while they stimulate the development and validation of the infrastructure to support permanent hu-

***Above:** In this artist's concept, two extra-vehicular astronauts have prepared an autonomous robot to embark on its mission of collecting samples from the Martian surface. Once the robot is released, the crewmembers located back at the base will have the ability to either remotely control the robot's activities or allow it to use its own logic systems to perform sample collection. Image courtesy NASA.*

man presence throughout the solar system." According to Huntress, the intermediate destinations "comprise a progressive approach to the long-term objective", the surface of Mars. "There is a logical progression to successively more difficult destinations," he explained. Each new destination has value on its own and is "scientifically and culturally compelling" in the words of the final report. And in one of the study's most innovative creative leaps, for each step, the development of only one fundamentally new type of space vehicle is required. "This approach requires incremental investments to maintain progress, rather than huge new budgets," he explained to the Senate committee. It would allow the program to exist under a relatively constant 'budget roof', not requiring peaks (and valleys) or roller-coaster funding.

Bite-size chunks

Buzz Aldrin, Apollo 11 moon walker, believes that breaking a "20-year space plan" into bite-sized chunks may also offer profound political advantages. "You have no idea how important I really believe [this approach] to be," he told me. Because space goals are set by whoever is in the White House, he believes that "we should design a series of four-year programs" as part of long-range strategy. From time to time, he continues, "you may slide some of the objectives due to delays, wars, politics – but you can't slide who makes the decisions and that's the President." Projects with well-defined short-term goals would be politically much more advantageous than one big project with a far-distant single goal.

How do we get there from here? How do we fulfill the fundamental truism expressed in the report, that the "critical first step in a long-term program of human exploration of the solar system is to finally get out of low-Earth orbit." And this is not to say that we would abandon that region, since the International Space Station will be operating for decades even after Space Shuttle missions complete its assembly about 2010-2011.

Going to L-2

The ISS can be useful for research and testing, but despite NASA's initial claims it cannot serve as a jumping-off point for more distant missions. "Its orbit inclination creates a severe penalty for station-launched missions to the Moon and planets," Huntress explained, referring to the sharp north-south flight path necessitated by making it accessible to Russia's spaceports. Combined with a crippling 25-30% performance penalty of US space shuttles launched into the station's inconvenient orbit, these considerations guarantee it will remain a space voyage dead end. In contrast, the IAA study focuses first on a region of empty space as the first human destination beyond LEO. It is located about 1.6m km from Earth (four times the distance to the Moon), away from the Sun and is designated "SEL2", or Sun-Earth Lagrange-2. It is a gravitationally 'neutral zone' where spacecraft are metaphorically swept along in the gravity wake of Earth and thus can maintain position there at very little cost in steering rocket firings.

Space astronomers have already had their eyes on this region. This is because of the unrestricted view of most of the celestial sphere and of the benign thermal conditions (no rapid day-night cycles as in LEO) and of the uniform gravitational forces that allow widely-scattered spacecraft to maintain very precise relative positioning. NASA's Webb Space Telescope will be located here, as well as its Constellation-X and terrestrial planet finder instruments. The European Space Agency is developing a range of observatories – named after scientists Herschel, Planck and Darwin – that are also to be deployed in this region.

Above: *Astronaut Edwin 'Buzz' Aldrin. In 1966, he and command pilot James Lovell flew the Gemini 12 spacecraft on a 4-day flight, which brought the Gemini program to a successful close. Aldrin also served as lunar module pilot for the first manned Moon landing, Apollo 11, in 1969. Aldrin followed Neil Armstrong onto the lunar surface, completing a 2hr 15min lunar EVA. Image courtesy NASA.*

These scientific instruments are all self-contained, unlike Hubble and they have been designed to operate untouched by human hands. But in years to come, as these pioneering SEL2 telescopes uncover new secrets of the Universe, the need may grow for human servicing and even human-assisted deployment and calibration for follow-on instruments such as those now being imagined to map the surfaces of extrasolar planets.

By lucky coincidence, this region also offers a unique view of the Earth. The angular size of Earth at this range is nearly the same as that of the Sun, providing properly positioned observatories with a continuous annular eclipse that backlights Earth's atmosphere whose profile can be monitored. And since objects in the zone are not really in orbit around Earth, they can transition from it into orbits around Earth with ANY desired inclinations, to access, inspect, repair or otherwise control any object in Earth-moon space.

Getting there

A human mission to SEL2 would involve launching a spacecraft just a little faster than the Apollo missions during the Moon race and would take about 15 days to get there. The velocity change, or delta-V – the propulsion maneuver – to stop in the zone and later to depart it for the return to Earth, would be about 20% greater than that needed by Apollo-8 in December 1968 to enter lunar orbit and then head back to Earth a day later. In hardware terms, this would take a rocket just a bit more powerful than Russia's 'Proton', perhaps a souped up version of the new generation of Delta or Atlas boosters now entering the US inventory. It would NOT need a massive Saturn-V-class behemoth from the Apollo days.

"What is being proposed is wonderful," noted Apollo 14 moon walker Dr. Edger Mitchell. "We need to be doing something like that – but we have to find enough excitement at any given mission to make it fun, to make it self-motivating." Although the main purpose of going to SEL2 would be just to acquire capabilities and knowledge, Mitchell told me it would be important to add enough 'sizzle' to make such a mission attractive.

The IAA plan calls for the development of a crew-carrying spacecraft with capabilities similar to those of an Apollo command module, or perhaps an upgraded Soyuz or Shenzhou vehicle. This capsule would end its mission by aerobraking in Earth's atmosphere. In many ways this vehicle resembles some proposed variations of NASA's new 'Crew Exploration Vehicle', or CEV – the designated successor to the space shuttle. The US would also have to build a reusable 'Service Module' for the propulsion and power. Between missions, this vehicle would be parked in low Earth orbit. There, it could be serviced and refueled. As mentioned, the step-by-step goals have been selected because each NEW step requires the development of only a single NEW spacecraft, along with evolutionary upgrades of vehicles already developed for earlier steps.

Furthermore, such a new spacecraft opens a multitude of doors. Once Earth builds a spacecraft for human access to SEL2, small modifications could also support missions to lunar orbit, or for servicing of constellations of communications satellites in 24-hour geosynchronous orbit – the velocity change and mission duration requirements are similar. It could also serve to support later human lunar landing missions, if this goal were selected.

Huntress explained to me why the IAA plan prefers a first step so different from any other strategy yet developed. "SEL2 only makes sense if you know you're going many times," he admitted," and it's NOT what you'd do if your only intent is to do it once or several times." NASA plans, in contrast, looked at basing in the Earth-Moon L1 point. "It's closer, it's in the line-of-sight of the Moon," Huntress explained, "but when you look at the energy requirements it is NOT the best place if you're going to go elsewhere."

Surprisingly, this IAA strategy does not require a lunar landing phase on the way to Mars. Huntress had testified that the Moon itself "is not necessarily on the critical path to Mars," and the draft report elaborated on this theme: "The Moon is a destination with important scientific and cultural benefits that make it worthy of human exploration," it stated, "but from a technical standpoint it is not necessarily in the critical path to Mars." Furthermore, it continued, "it is debatable to what degree the development of tools and habitats for the Moon will provide substantial benefits to eventual Mars exploration. . . . [T]hey require development of surface exploration capabilities that may be substantially different from those required for Mars." So there could well be good reasons for humans returning to the Moon, the report concludes – but preparation for Mars is not among them.

Gateway post to the asteroids

The term 'gateway' as applied to the SEL2 zone means that objects parked there – and it does take significant propulsive energy to get there – can trade that energy back in to be applied to pathways to other destinations. A vehicle could depart SEL2, dive back towards Earth and while swooping by it, fire its engine again to attain an extremely efficient escape trajectory. Because of this initial energy advantage, vehicles departing for more distant destinations would not face an excessively challenging velocity requirement. They would be easier to build, since much of the preliminary work would already have been accomplished just to get to SEL2.

For humanity's first sortie beyond the Earth-Moon system, Huntress and the IAA team had a consensus – visit a passing asteroid: "There is no doubt that a one-year human mission to a Near-Earth Object [NEO] would serve as an excellent intermediate step before any mission to Mars," he told the Senate Committee. The full report elaborated: "NEOs are ideally situated to provide an important stepping-stone to Mars. They are accessible with flight times that are intermediate between SEL2 and Mars and will provide us with an opportunity to exercise many of the required transportation elements in a relatively low-risk manner." Again, the notion that the next solar system object to receive human footprints could be a small asteroid is not new, but the study provides novel justification for it. Reaching an asteroid and spending several weeks exploring it (but not 'walking' on it – its gravity would be far too low) would provide a double bonus – a stepping stone for farther travel and an immediate object of intense scientific and practical interest.

There are powerful reasons for stand-alone interest in these particular objects. For planetary geology, determine if these are burnt-out comets with surviving residue of very old ice deep inside. For resource exploitation, find out if they contain materials susceptible to in-place industrial exploitation, from water to perhaps metal ores. And for the ultimate 'down home' justification, for threat mitigation, determine what is their typical consistency and internal structure and how they would it react to propulsive forces designed to alter their orbits.

"I find it a very refreshing approach," noted Apollo astronaut Rusty Schweickart. "I am especially supportive of their recognition of the critical role asteroids will likely play in our future in space." The incremental approach also appealed to him: "The step-by-step logical progression leading to real capability for human presence in deep space will also be more attractive to the public than one-shot grasp for a human Mars landing. Much of the required investigation can be done robotically," the IAA report concedes, "but it may ultimately be important to enable human explorers to use their powers of observation, intuition and active testing to fully understand the detailed physical nature of NEOs and to validate impact mitigation techniques."

An important feature of NEOs that argues for the advantage of moving directly to human on-site presence is the very long time between revisits. This is at first counter-intuitive - these objects are 'near' Earth, with orbital periods very similar to Earth's. But 'launch windows' only occur at intervals in which two objects 'lap' each other around the Sun – and if their periods are very similar, that 'overtake interval' can be a decade or more.

In practical terms, this feature offers little chance of robotic precursor missions to provide data that can be studied leisurely while planning a subsequent human visit. If there is a window for a robot mission, this means that the next window for astronauts could be many, many years in the future. So instead, once an asteroid is selected as a target for a human mission

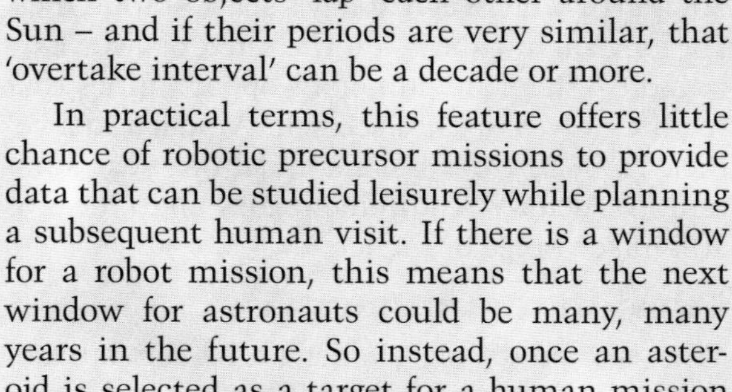

Above: *Astronaut Russell Schweickart. He served as lunar module pilot of Apollo 9 which was the first manned flight of the lunar module. He also served as backup commander for the first Skylab mission in 1973. After the Skylab program he transferred to NASA Headquarters in Washington, D.C.. Image courtesy Russell Schweickart.*

(based on observations from Earth), small probes could be dispatched on faster trajectories, to arrive a few weeks or months prior to a human crew already on its way.

Bridge to Mars?

"By far the strongest imperative for human missions to NEOs arises from consideration of their utility as an intermediate step to Mars," the report argues. "Their locations and physical characteristics will stretch the capabilities of human exploration just enough to greatly reduce the risk of the Mars missions to come. NEOs will thus play an important architectural role as a bridge between Earth's neighborhood and Mars." And later, if the choice at Mars is to first visit its moon Phobos, "a precursor mission to a near-Earth asteroid would allow demonstration of almost the entire mission at a destination closer to Earth, with ample solar power availability, high communications rates and relatively short return-to-Earth flight times that provide an extra measure of safety."

In plotting the paths from Earth to asteroids and back, one has to ask just how hard is it to get from 'here' to 'there'. It turns out that this questions depends on how you define 'here' (that is, where near Earth will you start from) and 'there' (which asteroid do you want to visit?). This IAA study provides a revolutionary definition of what 'here' means and thereby greatly facilitates the transportation challenge. Here's why: numerous studies have been made over the past three decades of round-trip missions to passing asteroids., with the total change of velocity (or delta-V as it is known to rocket scientists) and total flight time calculated. They all start in the familiar 'low earth orbit' where today's space station flies. For long flight times – on the order of two years or more – there are numerous opportunities that require only 6 to 7km/sec delta-V added to a vehicle in LEO. For missions under one year duration, the delta-V requirement goes up to the 9 to 10 km/sec range or higher.

Mission requirements

This is a pretty big challenge. In comparison, an Apollo lunar orbital mission required about half that much - 4.6km/sec delta-V (from both the Saturn S-IV-b upper stage and the Apollo's own Service Module). This Apollo figure compares to the delta-V capability of the space shuttle orbiter of only one tenth that amount, or the multi-spacecraft Apollo lunar landing and return delta-V of about twice that amount. A round-trip mission to Mars orbit (not landing) is estimated to require a little less than twice that amount (about 7km/sec). All of these high-velocity missions required multiple vehicles with very high fuel-to-payload rations – a daunting engineering and funding challenge. To underscore this, here's a 'typical' asteroid visit-and-return mission. The target would be asteroid 1991JW, allowing a 16 May, 2009 launch, five weeks at the object and a return after 363 days, all for a delta-V of 5.9km/sec. But figures in the 9 to 11km/sec range are much more common.

In the face of this formidable delta-V challenge, the strategy proposed in the IAA report is to stage such a mission from the SEL2 gateway. There, reusable space tugs initially developed for access to that point can deliver the asteroid-bound space vehicle and, in the end, its crew. The energy required to assemble the vehicles here is acquired piecemeal through reusable tugs over a period of months and is then expended efficiently during a brief launch maneuver in which the total required delta-V has already largely been 'pre-invested'.

As a result, in a typical 1-year mission (e.g. asteroid 1999AO10 in 2025), the additional delta-V is only 5.5km/sec from the SEL2 point. By changing the meaning of 'here' from low-Earth orbit right at the edge of Earth's deep gravity well) to the SEL2 gateway (way out on the 'lip' of Earth's gravity field), the 'next step' becomes much more manageable. The mission isn't free – this isn't trajectory magic – but the steps are much more 'bite sized'.

The interplanetary transfer vehicle

Since the flight time increases from a few weeks to many months, neither a modified Apollo vehicle nor the SEL2 human-access spacecraft could keep a crew alive long enough for an asteroid round trip. So the elegance of the IAA plan pays off again. Following the principle of 'one major new vehicle per step', it calls for development of an 'Interplanetary Transfer Vehicle' to carry a crew for a long mission (a year or more) with a delta-V capability of about 6km/sec (to be increased to 8 for later Mars missions). This spacecraft "is most significant development that will be required for this step," the report states. "This will represent a substantial investment and must be designed with the ultimate destination (Mars) in mind."

In terms of hardware, designers would be looking at a booster perhaps half the size of a S-IV-B upper stage from the Apollo era and a human spacecraft no heavier than the Apollo multi-module combo of Command/Service Module, Lunar Module. We know how to design and build something like this. The vehicle's engines will use tried-and-true chemical propulsion. "Although there are certainly other propulsion technologies that can provide better performance, at least on paper," the report explains, "in our view the risk and expense of those developments would only serve to further delay the first human journeys into the solar system." As new propulsion technology comes on line later, it would only enhance transportation capabilities already pioneered by these vehicles.

The report does not expect there to be only a few missions to a few asteroids and then a cessation of such exploration. The missions would likely continue and expand in scope and sophistication, independent of Mars activities. Different types of asteroids would be visited, so the full range of geology, mineralogy and internal structure could be assessed. But once the initial year-long asteroid sorties had validated the life-support hardware, the next big interplanetary step becomes feasible.

So, with the first two stepping stones now achieved – SEL2 with its fleet of science observatories and the variety of passing asteroids now also within reach – and with a Mars landing the ultimate goal, where next will human footprints be planted? Falling back upon the design principles of the IAA strategy, it is clear that the last step is still too great a leap. Too much remains to be designed and built and too much material (shelters and equipment and supplies) needs to be sent towards Mars, to be accomplished under the relatively constant level of funding and engineering challenges envisioned.

Cargo

So in the IAA plan, one other major space vehicle is needed even before the new human Mars landing craft itself can be built. This final intermediate step is to develop hardware to provide a means to get large cargoes to distant destinations where future astronauts can use them. As the report states, "a robust cargo delivery capability is a key element of a sustainable human exploration program." In more human terms, this means that techniques used by polar explorers a century ago now show the way to Mars – emplacing significant caches of supplies along the routes to be taken, so when the human travelers finally make the trip, they are not overburdened with life-critical baggage. "The principle of separating crew and cargo makes it possible to view the transportation challenge in a new way," the report explains. The key to doing this economically is to realize that inert cargo isn't in as much hurry as people.

This plan suggests developing high-efficient low-thrust engines – probably powered by nuclear reactors – to send most of the mission's cargo out ahead of the crew. This concept, again, isn't unique to the IAA study, but it is in keeping with the study's philosophy: "Only one major new capability is required for each step, coupled with evolutionary progress in existing capabilities. Our philosophy of incremental development as a means of managing cost and risk suggests that a human mission to one of Mars' moons, Phobos or Deimos, may be an important precursor to a mission to the planet's surface," the report continues. As the next stepping stone, this mission would be launched from SEL2 back near Earth, use crew transfer vehicles tested on asteroid missions and would rely on finding major components and supplies already waiting for the astronauts when they got near Mars.

Above: *This image, taken by the High Resolution Stereo Camera (HRSC) on board ESA's Mars Express spacecraft, is one of the highest-resolution pictures so far of the Martian moon, Phobos. The image shows the Mars-facing side of the moon, taken from a distance of less than 200km on 22 August 2004. Image courtesy ESA/DLR/FU Berlin (G. Neukum).*

To Phobos?

Dr. Fred Singer, now in his 80s, was a pioneer in the development of rocket and satellite technology and was the first director of the US Weather Satellite Service, now NOAA (Dept. of Commerce). Among his lifelong space science interests has been a passion for human missions to the martian moons. "I've personally been concerned with Phobos and Deimos for last 40 years," he told me. "The moons are not too difficult," he explained and "the main reason is because their landing velocities are zero. They are actual space stations in Mars orbit, you just tie up." His proposed research program, named 'PhD' (after PHobos-Deimos) pioneered the concept that it was worthwhile to make an interplanetary journey while not actually landing on Mars itself. Now the IAA strategy has endorsed this viewpoint.

Phobos has value both for future Martian missions (is there extractable water there? Can a radiation shelter be built using local regolith?) and intrinsically (what is its origin and when will it fall onto Mars?). And if a handful of 'martian meteorites' have been recovered on Earth, they must have left Mars in great numbers and many would have impacted on Phobos, for retrieval by explorers. Being close to Mars has many scientific advantages. "The removal of the light-time delay to Earth would make it feasible to actively manage experiments and react to discoveries," the IAA report points out, "thus helping to define the role of humans when they eventually reach the surface." A human-in-the-loop real-time control of the recent twin Mars rovers, for example, could have increased their surface speed by a factor of 50 and allowed the months of science investigations to be accomplished in a few days. Instead of creeping along at inches per hour and taking days to properly align instruments over rocks of interest, surface rovers (as well as flying vehicles) directly controlled by people on Phobos could operate at astonishing speeds and thus harvest even more astonishing results.

With refueling on Phobos, spacecraft based there can make sorties into lower orbits around Mars to rendezvous with robot payloads sent up from the surface with carefully-collected samples. Such samples can initially be studied and catalogued in the habitat imbedded safely under the radiation-shielding dirt of Phobos.

Astronaut James Lovell, who orbited the Moon on Apollo 8 and flew past it on the failed Apollo 13 landing mission, sees no psychological problems with an expedition that goes all the way TO Mars, but doesn't land ONTO Mars. "That would be well accepted by the public," he predicted and as for the crew, "their attitude would be fine. They would feel they were the pioneers who were the first to reach Mars," he speculated, based on his own crew's experience in circling the Moon without landing. "They would be elated, they'd be satisfied – as we were." Despite the long trip, he continued, they'd be happy to play a part in what he agreed was a logical strategy: "I concur – it has to be step by step," he told me, "you have to build on previous experience and each advance relies on past work." And the work of the humans who only orbit Mars will lay the groundwork for the next big step.

Colonel Buzz Aldrin, who was on the first mission to walk on the Moon, agrees. "I want to be strong about emphasizing that we should not be in a hurry to go the surface of Mars," he told me. "We can go to its moons," he continued, "and we should go down to its surface only when we're ready to work for permanent presence there." As an astronaut, he admits that "being that close to Mars could be frustrating," but if your primary mission is setting up a shelter on Phobos, it's plenty satisfying.

Above: *Astronaut Owen Garriott. His first space flight aboard Skylab in 1973 set (at the time) a new world record for duration of approximately 60 days, more than double the previous record. His second space flight was aboard Spacelab-1 in 1983, a multidisciplinary and international mission of 10 days. Image courtesy NASA.*

How soon?

How long will human visitors to Mars have to wait before going down to the surface? Some of the constraints of space travel give clues.

Opportunities to fly from Earth to Mars open up about every 26 months, the 'interplanetary window'. The window after the first human Mars orbit mission is too soon and probably the window after that, too, since the discoveries and the equipment operational experience will need to be studied in the fabrication of the next new space vehicle, the surface lander (and returner!). "Having completed exploration Steps 1, 2 and 3 prior to the first mission to the surface of Mars, a large suite of very capable hardware elements will have been developed," the IAA report states. "These will have been progressively evolved through each destination, so that by Step 4 the common elements should have the required capabilities. There will, however, be a large number of unique elements that are required for the Mars surface mission."

Funding their development and testing them – including space tests, perhaps near the Moon, perhaps even at Mars – will take a long time. It could be a decade or more between the arrival of humans at Mars and the first human footsteps ON Mars. The logic of this strategy may be persuasive, but as of now it still goes counter to 'conventional wisdom'. For many spaceflight theorists this 'stepping stone' approach is nothing but an overcautious temptation that distracts from the main goal, Mars itself. Instead, it replaces it with what could become a series of 'stumbling blocks' that would bankrupt a space budget and stretch out flight schedules beyond the attention spans of the public and political leadership.

"Besides", goes the main objection, "who could imagine going tens of millions of miles to look down on Mars only a few thousand miles below and then turn around and head back for Earth?" As it turns out, many people can – and their arguments in favor of such a strategy have gained both new urgency in light of the 'New Space Vision' from the White House and new credibility with the development of carefully designed and rigorously quantified strategies. "Some crewmember candidates will say 'If I spend years in preparation and then a couple of years in space, I must go to the surface to justify my time investment'," pioneering space station astronaut Owen Garriott has admitted. But he suspects that other potential crewmembers

Above: *Artist's concept depicting a possible scene when the first human travellers might walk on the surface of Mars. The area depicted is Noctis Labyrinthus in the Valles Marineris system of enormous canyons. The scene is just after sunrise, and on the canyon floor six kilometres below, early morning clouds can be seen. The astronaut depicted on the left might be a planetary geologist seeking to get a closer look at the stratigraphic details of the canyon walls. On the right, the geologist's companion is setting up a weather station to monitor Martian climatology. Image courtesy NASA.*

would not: "Other fully qualified candidates will be just as anxious to contribute to this magnificent opportunity to make a meaningful, early contribution to our exploration of the solar system and to the Mars vicinity," he continued. "This is a case in which 'self-selection' provides a quite valuable discriminator."

Garriott appealed to historical precedent, to which he was a direct eyewitness. "Consider selection of the early Apollo crews in the 1960s," he explained. "Probably everyone in the Astronaut Office would have wanted to go to the moon's surface. But some more than willingly accepted roles in lunar orbit or in LEO, or as back-up crews which were essential to the moon landings, or even reached for other important objectives, as in Skylab. I would expect Mars to be a similar situation."

And considering the length of time between the first human missions to the martian moons and the actual first footsteps on the martian surface, there might even be time for some of the early crews to return to Earth and spend a few years retraining for the landing mission itself.

Building capacity gradually

The detail of that final step is another topic entirely, since what is most innovative about the new strategic plan is how people get to that point. "This architecture gradually builds capability to explore the solar system through a series of carefully selected steps, each one designed to eventually enable humans to reach the Martian surface," the IAA study concludes. "This by no means implies that the first human mission to the planet will be easy. Even with the significant investments made in the earlier steps, this fourth and final step will be the most challenging and the time at which we will be ready to undertake it is uncertain. Ultimately it will be the continuing sense of exploration, along with the scientific discoveries and technical progress of the preceding steps, which will sustain public interest and international political support and make human presence on Mars a reality."

Looking back on the process of developing, critiquing and disseminating the report, Huntress told me that the peer review process had been surprisingly smooth. "We received about a dozen reviews from all over the planet," he explained, "and they were uniformly excellent." In 2003-2004, Huntress presented the strategy at international space conferences in Houston, Bremen, Paris and Vancouver, receiving "a uniformly good reaction." He also stayed in regular touch with the 'NASA Space Architect' whose responsibility it was to develop long range plans: "Our interactions with NASA were very, very good," he said. "Our study is unique," he continued, "in establishing goals and then deriving destinations", rather than past practices of picking destinations or favored hardware and then shaping a strategy around them. "Hopefully," he concluded, "our document will help."

Jim Oberg *was a mission controller and is now one of the world's best known writers on spaceflight. He unveiled many of the secrets of the Soviet space programme and now writes and broadcasts on American, Russian and Chinese space plans.*

NASA's Hubble Space Telescope acquired this picture of Mars on 28 October 2005, within a day of the planet's closest approach to Earth. The image shows a regional dust storm, nearly in the middle of the planet in this view, which is about 1,500km long measured diagonally – as big as Texas, Oklahoma, and New Mexico combined. Image courtesy NASA, ESA, The Hubble Heritage Team (STScI/AURA), J. Bell (Cornell University) and M. Wolff (Space Science Institute).

6

SOYUZ IN THE JUNGLE

Russia's main rocket was designed in 1953, flew in 1957 and has made over 1,700 launches. Here, *Laurent de Angelis* explains how, far from being retired, a new launch site is in preparation for this venerable rocket, in the South American jungle.

In a new international endeavour RUSSIA builds a launch base with Europe in South America

A RUSSIAN ROCKET launched from a French space base? Who would have thought that this was possible, just a few years ago?

There is nothing new about cooperation between France and Russia. In 1966, French president De Gaulle was the first westerner invited to visit and attend a launch in the then top secret Baikonour cosmodrome. But such cooperation was limited in many ways. Westerners visited Russia for missions and then returned. National rockets were always launched from their own national territories only. But this has now changed and what was previously impossible became possible. The Soyuz launch facility will require not only heavy permanent installations in French Guyana, but also the *permanent* presence of dozens of Russian engineers and technicians. Up to 200 people will come from Russia during each launch campaign.

The circumstances: politics and economics

Once the USSR came to an end in 1991, international cooperation on space topics could reach a new dimension. Once secret military and space technologies (this term had a very wide meaning in Soviet Union) were declassified. The military and space budget, which had benefited from comfortable subsidies, suffered a severe cut off after 1991. Many programmes were abandoned, such as the space shuttle Buran and its giant booster Energia, able to put 100 tonnes into orbit; and interplanetary probes projects. As a result, the space industry of the former Soviet Union had a crucial need of money. Alliances with foreign space operators appeared to be a good solution for most of them.

At the same time, on the western side, the main aerospace companies soon realized that some very good bargains could be made. They could obtain, for a very low price, unrivalled Russian space technologies, which had cost billions of roubles and had required millions of hours of work by the very best engineers of the former Soviet Union. Such alliances took various forms, from simple technology sales (e.g. Russian engines, build by NPO Energomash company and used on the Atlas V American rocket); western operators using Russian and Ukrainian rockets for many purposes (e.g. commercial launches). Alliances were made around the Proton rocket (Khrunichev and International Launch Services), the Zenit 3SL (NPO Yuzhnoye and Boeing) and, as we will see, Soyuz (TsSKB Progress and Starsem).

The idea of a new launch site

The Russians, for many years, wanted to have a launch site located on or near the equator, which gives a space rocket the best possible performance. They showed a great interest in the French Guyana space base and started to contact the French government, to talk about the possibility for them to build a launch complex there. In this particular case, things started in 1996, when the Aerospace Company EADS and Arianespace, the European space launch operator, created an alliance with the Russian Federal Space Agency and TsSKB Progress, which assembles the Soyuz rockets in Samara. Together, they created "Starsem", a joint venture company offering launch services for medium sized payload from Baikonour. The launcher used in this partnership was the famous Soyuz, which is based on a rocket dating to 1953.

The idea of using Kourou as a new launch base for Soyuz came out with a French-Russian study made by the Aérospatiale company (which became later part of the EADS company), in 1997, but not completed until 2003. Meantime, the Ariane 4 was abandoned, its last launch taking place in February 2003. That left only the Ariane 5 heavy class launcher, but a gap for a medium size launcher in Kourou.

The Soyuz project, initially French-Russian, became Euro-Russian; France proposed to the other European countries to join them, using the European Space Agency structure, (ESA).

The obstacles

Many obstacles had to be cleared before the project got really started. First of all, although the cold war was over, it took time for many people to get used to the new situation. Even if Kourou was a civilian space base, the base was never far away from sensitive technologies, such as military satellites and transmission equipment. For these reasons, some people where reluctant to accept the permanent installation of about two hundred Russians on the base. As we will see, this was eventually sorted out.

Opponents of the project argued that the Russian rocket would be a dangerous competitor for Ariane when installed in French Guyana. This argument lost most of its force when the Ariane 4 class was abandoned. That is because the Ariane 5 family, now used by Arianespace, is a heavy class booster, able to carry up to 21 tonnes of payload in low equatorial orbit and up to 10 tonnes to geostationary transfer orbit. Ariane 5 is comparable to the American Delta IV heavy. By contrast, Soyuz is a medium class rocket. Its expected performance from equatorial launch site is 11 tonnes in low Earth orbit or 2.8 tonnes to geostationary transfer orbit. Even with improvements, it will never be able to compete with the Ariane 5.

In addition, in case of a refusal, the Soyuz company would probably have found another site near the equator and the site of Christmas Island with Australia had been mentioned several times. It would become a competitor anyway. Putting Soyuz in Kourou would increase the base activity and the local economy.

Another problem was that the Russians had impressive technologies and knowledge - but no money. As a result, the basic principle of this project was that the Russian side will bring hardware, such as the rocket, platform blueprints and specific elements such as the arms supporting the rocket on the launching pad for instance, as well as technicians and engineers. The Europeans would bring: capital, their equatorial site of course and ensure the construction of the complex: access roads, buildings and platform.

Finally, the most difficult obstacle was the funding. The whole project represents a €344m. This is small compared to some major space programmes which cost billions of euros. The European Union tried to make this project a priority and it is indeed an European project, for six European countries participate (Germany, Italy, Spain, Belgium, Switzerland, France). But the European Union could not give its guarantee to the European Investment Bank for the €121m loan necessary to start the site. Finally, France alone decided to guarantee the loan and it was decided that this loan would be paid back by Arianespace with the profits generated by the launches. Thus, almost eight years after the preliminary study, the project had a firm financial foundation.

Several agreements will be firmed up between Russia, ESA and France to formalize the whole system. The Soyuz zone in Kourou, although managed by the Russians, remains French territory under French authority for access and safety rules.

The first jungle was cleared in July 2004 and bulldozers began work on the site itself in November 2004.

The French Guyana space base

The first French rocket to put a satellite in orbit, on 26th November, 1965, was not launched from French Guyana, but from Algeria, in the Sahara desert at a lonely place called Hammaguir. Algeria became an independent state in 1962, but a treaty left the Hammaguir base to France until 1967, when it returned to its traditional solitude. A new site must be found and after a lengthy selection process, the site of Kourou, in French Guyana, was finally chosen.

Kourou is located near the shore, on an almost flat place open to the Atlantic Ocean. This position allows a launch to any kind of orbit, from polar (heading north) to equatorial (eastbound). Whatever the trajectory, the rocket quickly flies over the ocean soon after lift off. Moreover, the latitude of Kourou, almost on the equator (5°18′ north, 52°50′ west), provides a maximum gain of power for geostationary satellites launched eastward. The angular speed of Earth, of course maximum at the equator, gives free energy. Compared to other space bases, a rocket launched from Kourou will carry 17% more payload than the same rocket launched from Kennedy Space Center in Florida (28.3° north). The difference is even greater if we compare it with the Baikonour cosmodrome in Kazakhstan, which is located at 45°, the same latitude as Bordeaux.

The French Guyana space base, which spreads over 90,000ha, is out of the hurricane zone. Winds, the Alizés, are moderate and blow in the same direction. The site is a not an earthquake area, for the underground, mainly precambrian rock, approximately 2bn years old, is extremely stable. The climate is equatorial and wet: French Guyana, except for a narrow strip of land along the coast, is totally covered by the Amazonian jungle on its 91,000 km² area.

The first launch from Kourou was performed in 1968 and the first space rocket, Diamant B, was launched in March 1970. Eight Diamant rockets and one Europa II carrying eleven satellites to low equatorial circular orbit and one dummy payload were launched between 1970 and 1975. There were three failures, including the single European Europa II launch.

The start of the Ariane program, with a first launch in 1979, marked the beginning of a substantial development of the French Guyana Space Base. Diamant was only a light launcher, weighing 24 tonnes at lift off in its last version, able to carry small satellites up to 120kg into low equatorial orbit. Ariane launchers grew in mass, power and performance, becoming comparable to the most powerful Russian or American launchers. Ariane 1 (1979-86) weighed 210 tonnes and could send up to 1.7 tonnes into geostationary orbit; Ariane 5 ECA weighs 780 tonnes and can carry 10 tonnes to geostationary orbit or 21 tonnes to low Earth orbit. Kourou now averaged up to twelve launches per year and there were even 14 launches in the 12 months between March 1995 and March 1996 and up to three launches in one month alone in December 1999.

Kourou now

From the crisis in the communication satellite business in 2001 and the end of Ariane 4 in 2003, the Kourou space base entered a transitional period, with a lower level of activity. From 2003, the number of launches dropped down from 10-12 to 3-6 per year. But two years from now, with the arrival of two new launchers, Soyuz and the light class rocket Vega, activity levels will increase significantly again.

The French Guyana Space Centre has six launching pads, as follows:

- ***Three launching pads are abandoned: probe rockets, Diamant and Ariane 4 (ELA-2)***
- ***One is currently working: ELA-3, for the Ariane 5. It is almost a double pad, for two platforms allow the preparation of two rockets simultaneously***
- ***One is being restored and modified: Ariane I-III, (former ELA-1) which will be used by the new Vega light class launcher, whose first flight is scheduled for 2007***
- ***One is under construction: The Soyuz Complex (named ELS: Ensemble de Lancement Soyuz)***

The Soyuz zone

The Soyuz rocket has distinctive features which make it difficult to reuse an existing platform, even if modified, such as the former Ariane 4 launching pad (ELA-2) for instance. Moreover, Ariane 5 carries many American satellites; and many American customers are reluctant to have a Russian neighbourhood close to their sensitive technologies. For these particular reasons, as well as other reasons such as the space available, topographic situation and safety, it was decided to

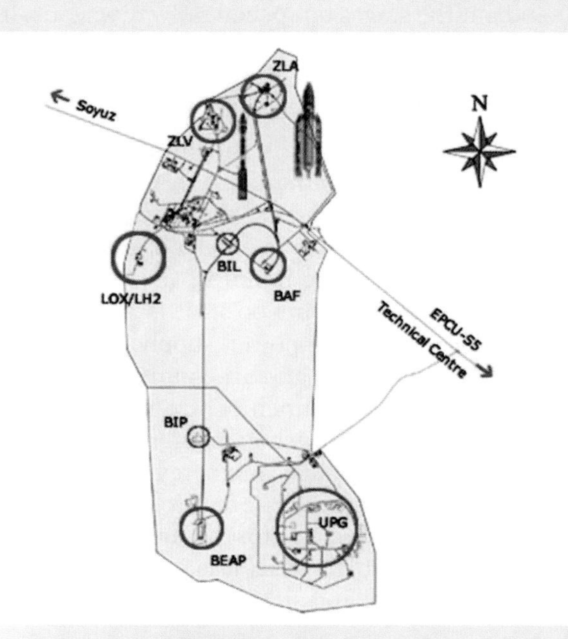

Left: *Map showing locations of facilities at the Kourou spaceport. From top: ZLA = ELA-3: Ariane 5 launch zone. ZLV = ELA-1: Vega launch zone. BIL = Launcher Integration Building. BAF = Final Assembly Building. LOX/LH2 = Facilities for producing liquid oxygen for the Ariane-5 main stage and the Vulcain engine, and liquid hydrogen for the Ariane-5 main cryogenic stage engine. BIP = Booster Integration Building. BEAP = Booster Engine Test Stand. UPG = Factory for manufacture of the propellants. The Soyuz launch zone (arrowed, top left) is located 13 km north-west of the Ariane launch site. Map courtesy of ESA/CNES/Arianespace.*

Top: An Ariane 5 GS launcher on the pad at Europe's Spaceport in Kourou prior to the successful launch of Europe's Meteosat Second Generation-2 satellite. Since Soyuz is a medium-class launcher, its performance will complement perfectly that of the ESA launchers Ariane and Vega, and enhance the competitiveness and flexibility of the exploitation of Ariane launchers on the commercial market. Image courtesy ESA/CNES/Arianespace-Service Optique CSG.

Above: The Soyuz FG launch vehicle carrying the Soyuz TMA-6 spacecraft is transferred to the launch pad at the Baikonur Cosmodrome, Kazakhstan, by rail in April 2005, in preparation for the Eneide Mission to the International Space Station (ISS) with Italian ESA astronaut Roberto Vittori. Image courtesy ESA-S. CORVAJA.

dedicate a zone approximately 10km away from the Ariane zone, now called the Soyuz zone.

The Soyuz zone is situated 27km from the Kourou Harbour and 70km away from Cayenne, on the territory of Sinamary commune. The Soyuz complex, which is about 2km^2 in extent, is divided into two distinct parts: a preparation zone and a launching zone. It is autonomous from the rest of the base: the platform has its own preparation facilities, safety installations, headquarters and countdown bunker.

The preparation zone

For safety reasons, such as an explosion on the launch pad during or shortly after lift off, the preparation zone is located 1,026m away from the pad itself. The vehicle assembly building is called MIK, an abbreviation for 'integration and test building hall' in Russian. The MIK will receive rockets parts for assembly, sent from the city of Samara in Russia to the harbour of St Petersburg and then on to French Guyana on a cargo ship. Once arrived at the harbour, they will finish their journey by road.

Two launchers can be stored and prepared simultaneously in this MIK building which is 56m long and 26m wide. It is only 20m high, which seems low for a 47m long rocket, but Russian rockets are assembled horizontally, not vertically like most Americans and Europeans rockets. They do not need a very tall building. The Soyuz is erected on its launching pad only a couple of days before launch. The MIK is equipped with two mobile bridges, able to lift up to 20 tonnes and put the stages on special railway wagons. It is connected to the launching zone using a 1,026m long railway. Technical rooms and laboratories are located next to the assembly hall itself.

The launch centre is the building where the launch sequence is managed. It is also the only building occupied during the launch phase. For this reason the place is shielded: a double concrete roof covers the building, able to protect it against a rocket falling on it. It is also equipped with an emergency power supply, able to give two hours of energy and air conditioning in case of problems. The launch centre is equipped for 200 people for a launch campaign. The preparation zone also includes an energy power unit, an air conditioning production unit and a fire building. Because Soyuz's kerosene fuel is not used by other rockets at the base, it is sent directly from Russia in tanks and stored on the site.

The launch zone

Launch pads differ a lot. Rockets using solid propellants usually stand on a very basic platform: a simple table with a hole under the rocket's nozzle place, for flames and hot gases exhaust during lift off and a simple mast, to provide electric supply and maintain the whole. Compared to this kind of platform, the Soyuz launching pad is very complex. The rocket must be filled with kerosene and liquid oxygen in each of its blocks - twelve independent tanks for the propellants, plus various tanks of other components (the upper stage, Fregat, is ready for use before it is put on the top of Soyuz). The system is not very automated, unlike later generation platforms and launchers and requires a large number of technicians working on the platform during preparation phase.

The launch platform is built according to blueprints furnished by the Russians. It has been adapted to the French fire and safety regulations. The platform is made of reinforced concrete lying on solid rock. It is 63.5m long, 43.3m wide and 16m deep. Under this level, a trench has been dug, approximately 30m deep. Its function is to take away flames and hot gases during lift off, for otherwise flames could damage the base of the launcher. The trench dug for the famous platform N° 1 in Baikonour, which is used for manned flights, is similar. A 15m diameter hole crosses the two upper levels to let the rocket get down and stand on the table. The bottom of the rocket in its final position is 7m lower than ground level.

The launching pad has six levels, of which five are underground. as follows.

- ***At the lowest level,* (more than 7m under level 0) is a mobile platform, built by the Russians, just above the trench. This allows the technicians to work and check the engines and the bottom of the rocket.**
- ***Level 0:* railway and lifting crane, kerosene and hydrogen peroxide feed lines, filling columns.**
- ***Level 2:* launch table, with four mobile arms. These arms (nicknamed 'the tulip') keep the whole in position until lift off. Almost unchanged since the very first flight, the system is ingenious. When engines are ignited and thrust has reached the full power (480 tonnes), a counterweight system causes the arms to swing and free the Soyuz.**
- ***Level 5.2* : communication systems, compressed air, valves, liquid oxygen feed line.**

Other levels: spare parts and technical rooms.

Above: *Artist's concepts of the Soyuz launch zone before and during a launch. The Soyuz launch zone (ZLS) will consist of three main areas: the launch platform where the launcher will be erected for liftoff; the preparation area (MIK) where the three stages will be horizontally assembled and controlled; and the launch control centre. The MIK is connected to the launch platform by a 1,026m railway which will be used to transport the launcher in a horizontal position. The launch control centre is located 1 km from the launch pad. Artist's concepts courtesy of ESA/CNES.*

A mobile launching tower 56.5m tall allows access to the various stages and to the satellites when the rocket stands in its final vertical position. The tower is used to install Fregat stage, the payload and fairing. Unlike the other stages, this block is installed in a vertical position. The mobile tower will protect the Soyuz and its payload from the wet equatorial climate of French Guyana until lift off. Four masts, each 80m tall, surround the launching pad to protect it from lightning.

The Soyuz rocket itself: basic characteristics

The original version of the Soyuz rocket is *R-7* or *Semiorka,* which means 'the seventh' in Russian. During the late forties and the beginning of the fifties, the Soviet Union built the R-1, R-2 and so on until the seventh type, R-7, built under the direction of Sergei Pavlovich Korolev (1907-1966). The letter 'R' means 'raketa', 'rocket' in Russian. The version we now call the Soyuz was first introduced in the 1960s. Soyuz can be compared to the DC-3 Dakota: still flying, when other planes from the same period now stand in the aeronautical & space museums. Over 1,700 launches have been performed since the first flight (see *Records*, chapter 1)

The Soyuz is a four stage machine, all using liquid propellants. It has a 311.5 tonnes mass at lift off and is 47.5m high. The first stage is made of four different blocks placed *around* the second stage. This parallel architecture gives a distinctive shape to the rocket. The first and second stages are ignited simultaneously. The four blocks are jettisoned after 118sec of propulsion, while the second works for 320sec. The first and second stage use the same propellants, liquid oxygen, stored at –183°C and kerosene and practically the same engines, known as RD-107 and RD-108. The RD-107 engine was developed by designer Valentin Petrovich Glushko (1908-1989) between 1954 and 1957. The RD-107 incorporates four main combustion chambers and two small steering chambers or vernier engines. Unlike western engines, the four main combustion chambers and nozzles are fed by a single turbopump (oxidizer pump and fuel pump on the same axel). The RD-108, powering the second stage, is based on the same design, except that its four nozzles have four vernier engines instead of two and that it works for a longer time. All together, 32 combustion chambers fire simultaneously at lift off ! The total thrust reaches 488 tonnes in the latest versions of the RD-107-108.

Soyuz 2 ST Fregat

The version to fly from Kourou is called the Soyuz 2 and the first flight of this improved model took place on 8th November, 2004 from the Plesetsk cosmodrome in northern Russia, near Archangelsk on the White Sea, using a dummy payload. This version, named Soyuz 2.1a, is equipped with a new digital calculator, more powerful than the previous model, allowing better control of the rocket. It has a much larger fairing of the Ariane 4 type, making the flight trajectory more delicate to control.

The third stage also uses a mixture of liquid oxygen and kerosene. For now, the third stage has had a RD-0110 four gimballed chambers engine, developed by the firm CADB Voronezh, providing 29.8 tonnes of thrust for 240sec. On the 2.1b version, it will have a RD-0124 engine, developed by the same firm and providing a comparable thrust of 30 tonnes in a vacuum, but a longer working time, 300sec and having a high specific impulse of 359 seconds (the specific impulse is the time, in seconds, for a kilogram of propellant to provide 1kg of thrust. A high specific impulse indicates high engine efficiency. Depending on the type of propellant used, it ranges from 240sec for chemical engines such as a solid rocket booster to over 460sec for the most efficient upper stages using liquid hydrogen and liquid oxygen).

The R-7 has used and uses at the present time many different types of upper stages, depending on the mission. The Fregat version will be used in Kourou. Fregat is built by NPO Lavochkin and uses storable hypergolic propellants (these are propellants that ignite on contact and do not require an ignition system. They can be kept at room temperature for long periods, but they are also extremely toxic). Fregat is a small stage, less than 1m high. It is made of six spheres containing propellant fluid. The Fregat's engine, the C5.92, can provide an adjustable thrust and can be re-ignited several times. A new enhanced version of the Fregat stage is under development.

The fairing has been developed by TsSKB Progress. It has a diameter of 4.11m and is 11.43m long. It weighs 1,700kg and is made of an aluminium bees nest structure, covered with a carbon fibre protection.

Mission profiles

The following mission profiles are envisaged:

- ***launch of satellite on geostationary transfer orbit, 36,000km above the Earth's surface, mainly telecommunication satellites***
- ***launch on a Sun Synchronous Orbit (SSO)***
- ***launch on Low Earth Orbit (LEO), or medium orbit***
- ***Interplanetary probes***
- ***Multiple payloads and micro-satellites.***

These missions required the Soyuz ST-Fregat at Kourou to have payloads from 5.5 to 11 tonnes. This means that the upper part – the Fregat upper stage, fairing and payload - weighs 13 tonnes.

First flights

The first flight, originally scheduled for 2006, is now expected for early 2008. The first payload is already known, the Optus B2 Australian communication satellite. The expected rate of launches of Soyuz from French Guyana is expected to be two or three a year at the beginning. It is not likely that Russia will launch national or governmental payloads from Kourou.

■ **Above:** *Artist's concepts of a Soyuz launcher blasting off from Europe's Spaceport at Kourou. The first launch will be an historic event, because it will be the first time that a Soyuz launcher lifts off from a spaceport other than Baikonour or Plesetsk. It will also be a milestone in the strategic cooperation that exists between Europe and Russia in the launcher's sector. The Soyuz launch vehicle that will be used at Kourou is the Soyuz-2 version called Soyuz-ST. This includes the Fregat upper stage and the ST fairing. Soyuz-2 is the last version in the renowned family of Russian launchers, and with improved performance will be able to place up to 3 tonnes into geostationary transfer orbit, compared to the 1.7 tonnes that can be launched from Baikonour in Kazakhstan. Artist's concepts courtesy of ESA/CNES.*

Communications satellite markets have been through a difficult number of years and some projects failed altogether, such as Iridium. The market began to recover from 2005. Soyuz is involved in launching the Galileo constellation of navigation satellites, which involves putting no less than 26 satellites into orbit. The first Galileo prototype was launched by a Soyuz Fregat from Baïkonour at the end of 2005.

If communications and other satellite markets continue to grow, the launch rate of Soyuz from Kourou could be significantly higher. The ESA council decision in December 2005 to give preference to the rockets being launched from the Kourou spaceport is of course good news for Soyuz in French Guyana.

Manned flight?

Manned flight is a possible development for this programme. The Soyuz launcher has shown an unrivalled reliability (Ariane 4 had impressive reliability with 74 consecutive successes, but it is no longer in use). Only minor modifications would be required to for such missions from the base.

Meantime, Russia decided to design a successor to the Soyuz capsule, the Kliper programme (see chapter 8). Kliper is a lifting body concept, able to carry up to six astronauts into orbit and has a mass of 17.8 tonnes. To lift this payload, which is more than twice the mass of the old Soyuz cabin, into low orbit, the old rocket needs another very important evolution: the Soyuz 3 version. Here, first and second stages would keep their original size, but the venerable RD-107/108 engines would be replaced. The third stage would be entirely new, a big stage using liquid hydrogen and oxygen with four engines giving 40 tonnes of thrust. Soyuz 3 would weigh approximately 392 tonnes at lift off. A first flight of this new model could take place around 2012. The question of using such a new version in Kourou is not, ultimately, a technical question, for it is a *political* question.

Laurent de Angelis *graduated from the Institut d'Etudes Politiques de Paris in 1982 and is a journalist with Le Point and Espace magazine. He is author of Guide to space launchers (1992, updated 2000).*

At the Baikonur cosmodrome in Kazakhstan, the Soyuz booster carrying the TMA-8 spacecraft is rolled out to the launch pad on 28 March 2006, for final pre-launch preparations. (Also opposite).

7
RETURN TO THE MOON

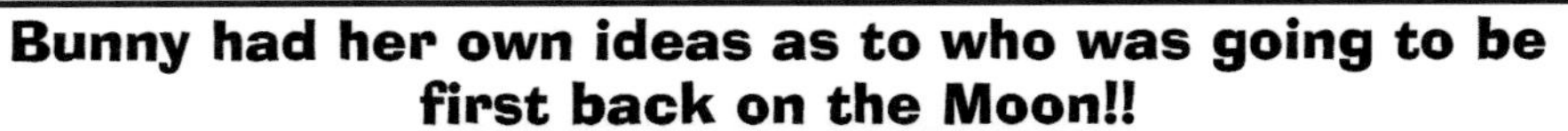

There was a long hiatus in lunar exploration following the end of the Apollo and Luna programmes in the 1970s. The 1990s saw missions resume with the Japanese Hiten and Hagoromo, the American Clementine and Lunar Prospector and Europe's SMART-1. With, under the Bush plan, the prospect of manned lunar landings before 2020, interest in the Moon is picking up again. This time, China and India will lead the return to the Moon, as *Paolo Ulivi* explains.

CHANG'E and CHANDRAYAN lead the way

SINCE THE 1990s, there has been a resurgence in studies of the Moon, which had seemed neglected by the spacefaring nations after the mid-1970s when the Soviet Union conducted its last lunar mission. Robotic lunar probes have been launched during the last one and a half decade by countries as diverse as the United States, Japan and the nations constituting the European Space Agency. These missions have provided new knowledge on our natural satellite, mapping it for the first time in different ranges of the electromagnetic spectrum to infer its composition, providing hints of its natural resources and controversial indications of whether water ice may exist in eternally shaded craters near the poles.

Two new countries have now emerged to make the next step in lunar exploration: China and India.

China's interest

China's space prorgramme was established by Tsien Hsue-Shen in the late-1950s. It is known that Chinese engineers reviewed the available open literature on the American Ranger lunar impactors during the 1960s. China's first satellite was launched in 1970 and ambitious objectives were drawn up for the space programme during the early 1970s, including manned flight. Once chairman Mao died and the gang of four was overthrown in the mid-1970s, the space programme was reorientated around applications and utilitarian tasks. In the words of communist party secretary Deng Xiaoping: "There is no need for us to go to the Moon and we should concentrate our resources on urgently needed and functional practical satellites".

For the next two decades and until the mid 1990s, the People's Republic carried out an extensive - if rather dull in the eyes of the layman - programme of application spaceflights, including telecommunication, weightlessness research, weather and Earth observation satellites, while studying navigation satellites. All major scientific programmes, including "space telescopes" were shelved.

Project 863

A new space programme was elaborated under the aegis of a technology research and development initiative called Project 863, introduced in 1986. One of its many stated objectives is to modernize space technology. The first ideas for a robotic lunar programme were developed under Project 863.

Preliminary mission studies were carried out at CALT (Chinese Academy of Launcher Technology) in the first half of the 1990s, their very existence being revealed in January 1995 by Jiang

Above: *Hiten Hagoromo: Japan's Moon probe, launched in January 1990. Hiten (originally called Muses-A) was an Earth orbiting satellite designed primarily to test and verify technologies for future lunar and planetary missions. The spacecraft carried a small satellite named Hagoromo which was released in the vicinity of the Moon. Hiten itself was put into a highly elliptical Earth orbit which passed by the Moon ten times during the mission and then in lunar orbit when Hiten was intentionally crashed into the Moon on 10 April 1993. Image courtesy JAXA.*

Jingshan, manager of the space programmes at the Academy of Sciences, who announced that the next five-year plan would include deep space and lunar missions. This was not to be, however, but in the following years planning and technology development progressed, lunar missions being the subject of at least three scientific and engineering workshops held at CALT and at the Qinghua technical university in Beijing.

An exploration plan in three phases was eventually presented to the central government immediately after the 2002 communist party congress and was publicly announced on 28 February 2003, when approval was obtained.

Chang'e

Named Chang'e (pronounced in English 'Tchang -er') after the fairy who flew to the Moon in a traditional Chinese legend and in fact one of the oldest "moonflight" legends known of all human cultures, the plan would start within a few years with an initial orbital reconnaissance mission, would continue later in the decade with soft landers, to carry automatic rovers for enhanced exploration at a later stage and would reach its climax around 2020 with the launch of automatic sample return missions. Like the Shenzhou programme of manned spaceflight, which reuses much Russian hardware and expertise, the lunar programme would see extensive help from the Russian space industry, to the point that joint missions could also be flown at some time.

In the ambitions of Chinese space planners, the Chang'e programme could then be followed around 2030 by human missions, although non-Chinese sources point out that simple human lunar circumnavigation missions similar to the Soviet Zond planned for the late 1960s could be staged in a matter of years reusing Shenzhou technology and the new heavy weight launcher to debut later in this decade.

Having developed a whole panoply of launchers for its application satellites and in particular high performance rockets to place heavy communication satellites in geostationary orbit, China has no lack of available launchers for its lunar

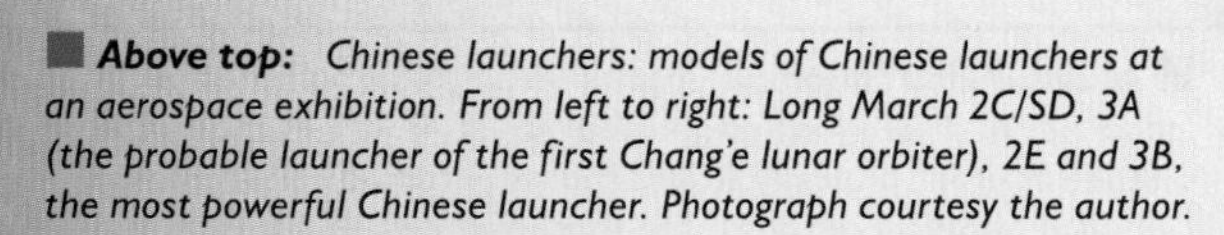

Above top: *Chinese launchers: models of Chinese launchers at an aerospace exhibition. From left to right: Long March 2C/SD, 3A (the probable launcher of the first Chang'e lunar orbiter), 2E and 3B, the most powerful Chinese launcher. Photograph courtesy the author.*

Right: *Long March 5: a model of the new Chinese heavy launcher, sometime called the Long March 5. Its introduction, during the 2010s, will make possible human lunar circumnavigation missions and robotic sample retrievals. Image courtesy of Patrick Roger-Ravily.*

missions. Small probes under 1,000kg could be launched on derivatives of the CZ-2 (Chang Zheng, Long March) rocket, a derivative of which is also used to loft the Shenzhou spaceships. For heavier spacecraft, the CZ-3 family would be used, whose performance toward the Moon ranges from 1,700 to 3,400kg. In fact, the first

Chang'e orbiter will probably fly on the CZ-3A, a three-staged rocket with a high performance third stage using liquid hydrogen and liquid oxygen as propellants and including engine technology equivalent to that of the European Ariane. During the next five to ten years, the debut of a new generation family of launchers, unofficially designated CZ-5, is anticipated. Depending on different arrangements of three "core stages" and boosters, the CZ-5 could launch between 4 and 10 tonnes to the Moon, thus enabling the sample return missions, Shenzhou lunar circumnavigation and more advanced deep space probes.

Mission profile

According to Chinese sources, the first Chang'e probe, expected around 2007, will be based on the DFH-3 (Dong Fang Hong-3)communications satellite, a large geostationary satellite bus which has already proved successful in several flights since its 1994 debut.

For the lunar mission, the DFH-3 will be equipped with a lunar orbit insertion stage, similar to the "smart" stage already used to place in orbit small communication satellites. This stage will double as an independent spacecraft and after taking the probe from an eccentric Earth orbit into a lunar orbit 200km high, it will manoeuvre to impact the Moon, at the same time when Chang'e will be flying above, collecting images and spectra of the cloud of dust raised by the impact to measure the composition of the upper layers of the lunar surface.

During its orbital mission, the 2,300kg Chang'e orbiter will take stereo pictures of the lunar surface while also analyzing the distribution of chemical elements and estimating their abundance. It will also survey the thickness of the Moon's dust layer and will explore the deep space environment between the Moon and the Earth. To meet these objectives, the probe carries six instrument. A digital camera will take stereoscopic images of the Moon using three CCD sensors, one looking directly under the probe, one looking forward and one looking backward. The camera will take relatively low resolution images showing details down to 120m in size. A three dimensional model of the Moon will then be built by integrating these images with a map of the lunar relief obtained by an altimeter which fires a laser beam at the Moon, the travel time of the pulse giving information on the distance between the orbiter and the ground below. An imaging spectrometer will provide images of the Moon that could be color-keyed to reflect differences in the composition of the surface layers. Other data on the distribution and abundance of elements will be obtained by gamma-ray and X-ray spectrometers, which detect radiation emitted by substances on the surface of the Moon.

Helium 3?

Chinese scientists openly express hopes that these instruments will be able to map the deposits of helium-3 (He-3) which are believed to exist on the Moon since it was detected in Apollo samples. He-3 is expected to provide the cleanest and most effective fuel for fusion power

Above: *Dong Fang Hong: a model of the Chinese Dong Fang Hong (DFH) 3 communication satellite, which will form the basis of the Chang'e lunar orbiter Photograph courtesy the author.*

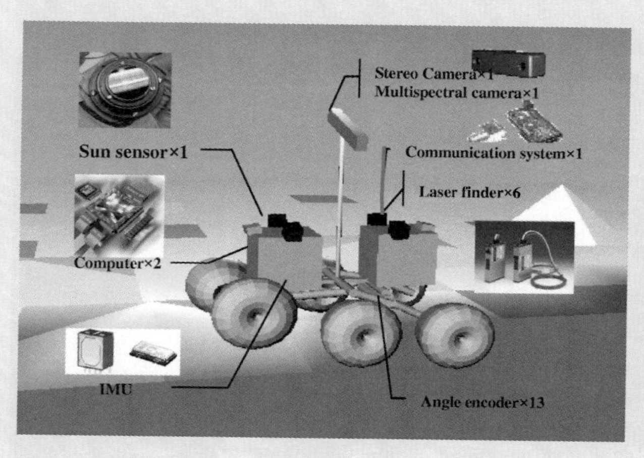

stations that may provide most of the Earth's energy needs by the mid of this century. The payload is completed by an instrument which measures the temperature of the surface, from which information on the thickness and nature of the dust layer can be inferred and by two instruments to detect solar particles.

Although most information that will be returned by Chang'e-1 has already been collected in some way or another by other probes, some of its objectives are unique and in particular the often-stated interest for searching for He-3. Laboratories have been built for the task of developing the instruments and most have been reported as being ready during 2005 with no major technical problems. A deep space communication network is being built, including a 50m steerable antenna built near Beijing.

After Chang'e

The second step of the Chang'e programme, after one or two orbiters have flown, will see a Chinese probe landing on the Moon. A few designs have been revealed, ranging from airbag-protected "hard landers" to small conventional "legged" landers probably incorporating some degree of artificial intelligence for controlling the landing manoeuvre itself.

Another very active field of study of Chinese space engineers is robots and rovers. Many designs with different degrees of autonomy have been revealed by the researchers of the many Chinese universities, institutes, laboratories and other institutions working on lunar rovers, one of which is said to provide the actual hardware for the space mission after a nationwide competition. They range from six-wheeled designs incorporating the "rocker-bogie" suspension system of the US Spirit and Opportunity Mars rovers, to articulated-body designs similar to old Russian Mars rover designs to unconventional two-wheeled or tracked robots. The rovers will be equipped with stereoscopic cameras to take navigation pictures along the way, as well as scientific images and panoramas and will probably carry robotic arms to collect rocks and other samples or to bring instruments in contact with the ground to conduct chemical and morphologic analyses.

Above: *Six wheel vehicle: a computer-graphics rendering of a possible six-wheeled Chinese planetary rover. Image courtesy of Ju He-Hua.*

Automatic sample return missions, the last step of the Chang'e programme, will probably also use rovers or robotic arms to choose and pick rocks and other samples. The four-tonne sample return probe is the only Chang'e mission that could not be flown on existing Chinese launchers and will have to wait until the development of the "CZ-5" is complete. Repeating a feat that only the Soviet Union has attempted in the past, sample return probes will probably look similar to Soviet Luna spacecraft, with a landing stage, an ascent stage and a simple spherical sample return capsule. However, these missions will not fly for the next fifteen years and therefore their details are not yet firm.

Beyond the Moon?

Also in China's space plans is the start of deep space and planetary exploration. Models of a Mars orbiter and of a Mars rover have been shown at domestic congresses and exhibitions and studies have been made of Mars landing techniques. However, the main interest of Chinese scientists and engineers, after the Moon, appears to be the exploration of the Near Earth Objects (NEOs). Studies of a small probe, called Expedition 1 and resembling the US Near Earth Asteroid Rendezvous (NEAR), which spent one year orbiting one of the largest NEOs, (433) Eros, finally landing on it, have been presented in several occasions at international astronautics congresses. Expedition 1 could rendezvous with a kilometer sized asteroid, either (1627) Ivar or (4660) Nereus, the latter being one of the easiest deep space targets to reach beyond the Moon.

China has also expressed interest in staging a mission that would soft-land a craft to test an engine capable of pushing a comet or asteroid off collision course with Earth. Engineers involved in these studies would like Expedition 1 to fly before 2010, but it is not known what support, if any, it is receiving from Chinese space officials. It is clear, however, that a NEO mission would be a fitting second step in deep space exploration after the Chang'e lunar orbiter.

India's interest

Asia's second most populous country after China, India, is also making plans for lunar exploration. After conducting an extensive programme of sounding rockets, India finally succeeded in launching a small satellite with its own indigenous launcher in 1980. India did not have very powerful rockets at first and therefore it conducted for a decade a series of launches of small satellites, often carrying astronomy payloads. At the same time, it relied on the Soviet Union and Europe to launch its larger Earth resource and communication payloads.

When it finally managed to develop heavy duty launchers, in the 1990s, India started to make plans for a more ambitious space programme, to include deep space and planetary missions. Human missions could also be flown at some time but are not yet approved, although prototypes of a recoverable capsule, which could form the basis of a manned spaceship have already been tested.

After years of intense studies on the possibility of flying a small lunar orbiter, go-ahead for the mission was announced by prime minister Atal Bihari Vajpayee himself during his 15 August 2003 Indian independence day address.

Chandrayan and its launcher

The Chandrayaan (lunar probe) spacecraft, based on the Indian Resource Satelite remote sensing spacecraft family, will have a launch mass in excess of 1,000kg, 440kg for the spacecraft proper and scientific instruments, the rest being fuel for the orbit insertion and for maintaining a 100km altitude above the Moon during the two year mission.

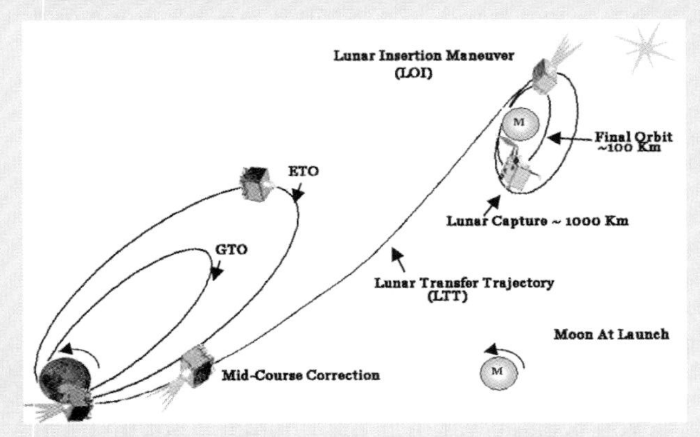

India at present has two rockets available for launching satellites: the PSLV (Polar Satellite Launch Vehicle) and its derivative, the GSLV (Geostationary Satellite Launch Vehicle). With a career spanning more than a decade, the PSLV is the preferred choice for Chandrayaan, which (just like Chang'e) will be inserted in an eccentric Earth orbit first, before entering a lunar transfer trajectory that will take it to within 1,000km of the Moon. After entering a high capture orbit, Chandrayaan will descend to the operational circular polar orbit. This trajectory has already been tested by the PSLV, up to injection in the transfer orbit. The more powerful GSLV could place a more massive spacecraft in lunar orbit and may be used for follow-on lunar missions or for future deep space missions to the near planets, asteroids and comets. It is worth noting that Indian missions to Venus, to Mars and even an orbiter to Mercury using the GSLV were under study in the early 1990s.

Mission objectives

The scientific objectives of the Chandrayaan mission will be quite similar to those of Chang'e: obtaining a three-dimensional map of the Moon and its mineralogy and composition. Moreover, it will continue studies on the presence of water at the poles. Both US lunar missions of the 1990s, Clementine and Lunar Prospector, have made observations hinting at the presence of large quantities of ice in permanently shadowed craters at both the north and south poles. It is assumed that this ice was either carried to the Moon by cometary impacts, or it could have formed by chemical reaction of lunar minerals with hydrogen in the solar wind. A firm proof of the presence of ice would constitute a big boost for future crewed missions to the Moon, for it could be used to obtain water for astronauts and fuel for rockets. Despite the hints, however, the presence of ice is still disputed by many scientists and alternative explanations have been put forward that would explain the observations of Clementine and Lunar Prospector just as well, without implying the presence of ice.

Chandrayaan is expected to make detailed observations of several selected sites, instead of making "global" observations, like Chang'e. It will map the geology and mineralogy of some young craters, including the Davy Catena, a remarkable string of craterlets probably created by

■ ***Left:*** *Path to the Moon: the mission profile of the Indian Chandrayaan lunar orbiter. That of the Chinese Chang'e probe is similar. Image courtesy of Narendra Bhandari, and reproduced with permission from Current Science.*

Germany and based on the experience of the European SMART probe, while Bulgaria will have a small instrument for monitoring the radiation dose in lunar orbit, which could yield interesting information for astronauts' flights to the Moon. American researchers will fly another mineralogy mapper and an interesting miniature imaging radar that will be used to map the permanently shadowed polar landscape, that would otherwise be unobservable to conventional cameras.

Another remarkable difference between the Indian and Chinese lunar policies is that India at this time does not have a clearly stated prospect for follow-on missions, while the Chang'e programme envisages a series of incremental steps stopping just short of human missions. It is possible that Indian lunar landers may follow, which could even include a small GSLV-launched sample return mission.

the impact of the fragments of a small crushed comet. Moreover, it will dedicate detailed observations to the Marius region of volcanic hills, a site proposed time and again for an Apollo landing but always rejected on the end. Observations will also be dedicated to the South Pole-Aitken basin, one of the largest craters of the solar system which may have excavated deep into the lunar crust, bringing vast portions of the lower crust and mantle to light.

Payload

The basic scientific payload is also very similar, if not identical to Chang'e, consisting of a stereoscopic camera (capable of "resolving" details only 5m across); an imaging spectrometer; three X-ray and gamma-ray spectrometers and a laser altimeter. Just like Chang'e, moreover, Chandrayaan may carry a small 25kg impactor to crash on the Moon during the mission under the eyes of the mothership. Unlike the Chinese probe, however, which will be a mostly domestic enterprise, the Indian probe will have ample room for international payloads and contributions.

Replies to an Indian tender for international instruments to be mounted on Chandrayaan were received from all over the world including ESA, the US, Russia, Germany, Bulgaria etc. Israel was said to have proposed including a small national subsatellite to be released in lunar orbit, carrying some kind of telescope. In the end, ESA will fly two instruments for resource mapping, one built in England, the other in

Above: *Artist's impression of SELENE (SELenological and ENgineering Explorer), a Japanese Moon probe, planned for launch on-board an H-IIA launch vehicle from Tanegashima Space Center in 2007. SELENE consists of a main orbiting satellite at about 100km altitude and two small satellites (Relay Satellite and VRAD Satellite) in polar orbit. Image courtesy JAXA.*

Asian Moon race?

When the lunar projects of both China and India were kicked off, there were critics, both internal and international, arguing that the countries were embarking on pointless endeavors and that money spent on deep space exploration could be spent in better ways even inside the space programmes themselves. However, lunar plans at present constitute only a minor part of the space agencies' budgets. In the case of China, the Chang'e orbiter is reportedly costing a meager (by space standards) $150m, to be compared with more than $2bn spent on the development and first flights of the Shenzhou programme. Critics also noted that neither country had real planetary science laboratories and institutes and it is remarkable to see how quick they were established virtually from scratch.

There have been suggestions that Chang'e and Chandrayaan might spark a second "Moon race" after that of the 1960s between the United States and the Soviet Union, more so as the two Asian countries are also regional rivals and have even been at war over their borders. But up until now there are no signs of one such race and the two space agencies have even established cooperation in some fields. If an "Asian space race" ever materializes, it now seems increasingly likely that it will be between China and Japan, whose Asian space primacy pride appears badly hit by the Chinese manned flight ambitions.

The return to the Moon

China and India will not be the only countries that will send automatic probes to the Moon in the next years. A European probe, SMART 1, entered lunar orbit in 2004 and it is addressing some of the same objectives of Chang'e and Chandrayaan. Japan will launch its heavy SELENE orbiter sometime in 2007. Being the heaviest and best instrumented lunar orbiter launched in many decades and carrying also two small subsatellites, Selene is slated to make extensive observations and discoveries of lunar geology and astronomy. Japan could also launch its much delayed Lunar A probe before the end of the decade, although its original launch date was to be 1997. The probe, whose design has been changed many times due to a long series of technical problems, will deliver two "penetrator" hard landers to the surface of the Moon for seismic studies that promise to be much more accurate than those carried out by seismometers left on the Moon by Apollo astronauts. Finally, the United States will also fly a robotic mission, the Lunar Reconnaissance Orbiter (LRO), sometime after 2008.

LRO is an orbital pathfinder to the manned missions expected to be flown after 2018 under President Bush's "Vision for Space Exploration" programme and will perform a year-long mission in a very low lunar orbit to map the surface and its natural resources. Its payload will again be very similar to that of both Chang'e and Chandrayaan, although its camera will be able to resolve objects as small "as a beachball" and may even put to rest for good rumors of the "Apollo lunar landing hoax", being able to show objects and vehicles left on the Moon by the astronauts. The main emphasis of the LRO mission, however, will be observing the poorly known polar regions, where it will again investigate the presence of ice, in support of a permanent manned base that could be built there. For this task, it will carry a completely new instrument especially designed to detect surface water ice deposits and frost, providing ultraviolet images of the permanently shadowed craters at the poles, lit only by faint starlight. After LRO, NASA's Robotic Lunar Exploration Program (RLEP) is to fly a next generation lander, which could be launched as early as 2010, to demonstrate a capacity for precision landings at targeted locations on the Moon; evaluate landing site environment; and determine if lunar resources can support a sustained human presence.

Luna 24 remains the last probe to soft land on the Moon. Which of these Moon probes will be next?

Paolo Ulivi was born in Florence, Italy and is a graduate of the Politecnico of Milan University in aerospace engineering with a thesis in the robotics of a Mars sample return mission. He currently works in aerospace and railway engineering and in his free time is an amateur astronomer, specializing in solar system observations.

Above Left: *Artist's impression of the Lunar Reconnaissance Orbiter (LRO), NASA's next Moon probe. LRO is the first of the Robotic Lunar Exploration (RLE) missions, and planned for launch by late 2008, it will orbit the Moon nominally for one year. Image courtesy NASA.*

8

In the footsteps of SOYUZ

When the first Soyuz was launched in 1967, few can have imagined that it would still be flying almost forty years later. But is a new manned Russian spacecraft at last within sight? Bart Hendricx has the details.

Russia's KLIPER spacecraft

FOR NEARLY 40 years, the Soyuz spacecraft has been the only manned vehicle to fly from Soviet/Russian soil. Although the spacecraft has undergone numerous modifications over that period, its basic design and capabilities have remained unchanged. Until early 2004 it looked as if the latest version, Soyuz TMA and its further modifications were going to serve the Russian space programme indefinitely. That picture changed when Russian Space Agency chief Yuriy Koptev held a news conference in Moscow on 17 February 2004. Answering a totally unrelated question about possible manned Soyuz missions from Kourou, Koptev remarked almost in passing that RKK Energiya, the country's leading space company, was working on a new manned vehicle "with a reusable return capsule and with a mass of 12-14 tonnes". Two days after Koptev's disclosure, Nikolai Bryukhanov, the deputy general director of RKK Energiya, revealed to journalists of the Russian space magazine *Novosti Kosmonavtiki* that the vehicle was called "Kliper" ("Clipper") and that Energiya had begun work on it "on its own initiative" as early as 2000.

Clearly, the project is still in the conceptual stage, but even in the two years or so that have elapsed since it was unveiled it has undergone a number of changes. At this point it remains uncertain if Kliper will be a lifting body or winged vehicle and which launch vehicle will be used to place it into orbit.

Origins

When RKK Energiya began working in earnest on a Soyuz successor in 2000, three configurations were studied for the re-entry vehicle: a capsule-type vehicle, a winged vehicle or a lifting body. While nothing is known about the first two designs, it would appear that the idea of developing a lifting body class vehicle as a successor to Soyuz evolved from research conducted in the late 1980s and early 1990s on a variety of re-entry vehicles intended to return cargo from space stations and perform microgravity experiments in orbit. This work began in 1988, with the development of the Raduga ballistic capsule which returned cargo from the Mir space station. Launched aboard Progress spacecraft, it could be loaded with about 150kg of cargo and separated from Progress just before the latter burned up on re-entry. Nine such capsules were flown between 1990 and 1994.

The next planned step was to turn these capsules into autonomously functioning spacecraft by equipping them with their own instrument/propulsion sections. Called Small Orbital Spacecraft (OMA), they would fly solo unmanned missions to carry out various experiments in microgravity and return the results to Earth. Three such OMA vehicles appeared on the drawing boards, two using the Raduga capsules and the Kosmos-3M booster (return capacity 150kg) and one employing an enlarged version of Raduga and the Tsiklon-2 booster (return capacity 450kg). NPO Energiya proposed an OMA for the German Express spacecraft, but this was turned down in favour of a competing proposal from the Salyut Design Bureau.

Eventually, the goal was to replace the ballistic capsules by lifting body re-entry vehicles known as Recoverable Manoeuvrable Capsules (VMK). These would have a cross-range capability of 1,000km and be capable of making controlled re-entries, resulting in reduced g-forces and landings with an accuracy of about 2.5km. The VMK was to land vertically using parachutes and soft-landing engines. The idea was to have four VMKs of different sizes attached to individually tailored instrument/propulsion compartments. The launch vehicles for the four versions were Tsiklon-2, Soyuz, Zenit and Proton. The Tsiklon-launched version (known as the Transport Research Vehicle or TIK) was apparently designed for unmanned microgravity missions similar to those to be carried out by the OMAs.

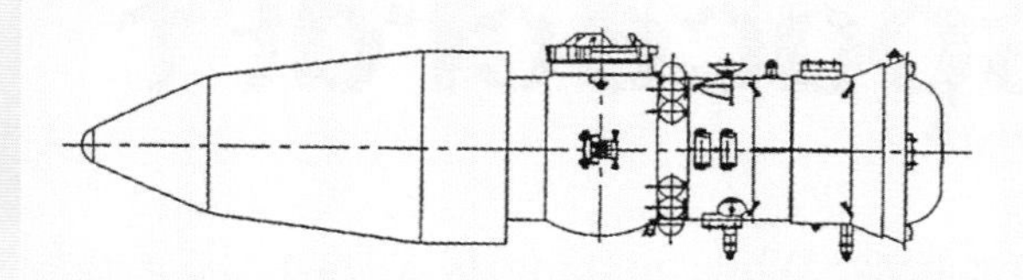

The VMKs were also said to be designed to return large and heavy payloads from space stations, with the three heavier versions also capable of acting as space station lifeboats. The manned versions were reportedly called Piloted Transport Spaceships (TPKA). A drawing of the Zenit-launched version shows a re-entry vehicle shaped almost exactly like that of Kliper and a docking compartment with a north-facing androgynous docking port inbetween the re-entry vehicle and the instrument/propulsion compartment. A patent application for this type of re-entry vehicle was submitted in July 1994.

Indications are that the VMK studies were discontinued in the mid-1990s, but then were picked up again when RKK Energiya began working on a Soyuz successor in 2000. Patents for other lifting bodies were granted to RKK Energiya after 2000 and may have been in the running for Kliper as well. One concept consisted of a lifting-body type re-entry vehicle and an instrument/propulsion compartment. The docking collar was on the re-entry vehicle itself and therefore probably reusable. There are no indications in the patent as to how many crew members this vehicle could carry [1].

Above: *Zenit-launched manned spacecraft with lifting body re-entry vehicle proposed in the early 1990s. Image courtesy RKK Energiya.*

Right: *Exploded view of Kliper. Image courtesy RKK Energiya.*
Key : 1 - fuselage shaped as lifting body
2 - winged fuselage
3 - crew cabin
4 - body of the Aggregate Compartment
5 - habitation compartment
6 - launch vehicle adapter with emergency escape rockets

Design

In early 2002, RKK Energiya decided that Kliper would use a lifting body type re-entry vehicle and it would seem that that by this time the VMK type design had won the day. However, during a press tour of RKK Energiya facilities on 30 November 2004, company officials revealed an alternative winged design for the fuselage and a choice between the two remains to be made at the time of writing.

The present Kliper has a reusable "Return Vehicle" (VA), made up of a crew cabin ensconced in an unpressurized fuselage which could be either a lifting body or a winged design. Attached to the aft of that is the expendable "Aggregate Compartment" (AO), consisting of a habitation compartment based on the Soyuz orbital module surrounded by the AO body. About half of the habitation compartment protrudes from Kliper's aft section and has a docking port to link up with the ISS or other spacecraft.

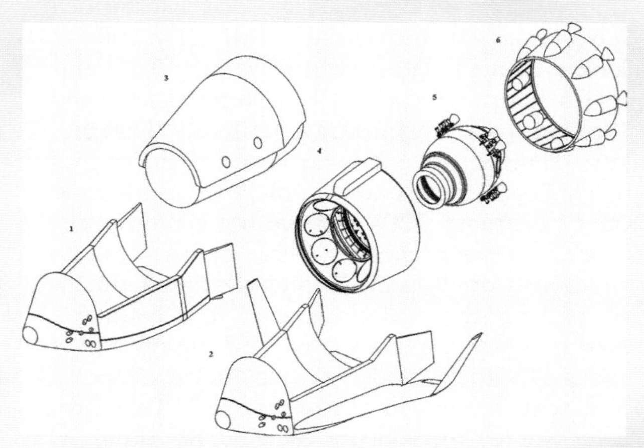

In short, Kliper's design cannot be traced back to one single vehicle, but reflects the experience accumulated over the years by RKK Energiya in designing capsule type spacecraft, winged vehicles and lifting bodies.

Return Vehicle

The Return Vehicle's blunt-shaped crew cabin offers 20m^3 of working space (five times as much as Soyuz) and can house a maximum crew of six (minimum crew two). The earliest cut-away drawings showed three seats in the front and three in the back, but in the latest design there are just two seats in the front (for commander and flight engineer) and four in the back (for passengers). The seats can be reclined to provide more comfortable conditions for the crew during re-entry. The crew cabin contains the most valuable equipment (control systems, life support, thermal control). The life support systems are largely based on those used on Soyuz-TMA and the Russian segment of the International Space Station. They can sustain six cosmonauts for five days during space station missions and for fifteen days during autonomous missions. The cabin has a side hatch for crew entry and exit and also sports four windows, two on either side.

The independently developed fuselage protects the front and lower part of the crew cabin during re-entry, descent and touchdown. The fuselage and the upper part of the crew cabin will have a heatshield consisting of 60 x 60cm thermal covers made from the same material as Buran's tiles. This should make it possible to reuse the Return Vehicle up to 25 times. The fuselage's nosecap will be covered with a non-reusable ablative heatshield similar to the one used on Soyuz. Installed in the fuselage are sixteen 24kg thrust attitude control thrusters and associated propellant tanks (containing liquid oxygen and ethanol). Twelve are located in the front section (six on either side) and four in the aft section (two on either side).

Behind the nosecap will be liquid oxygen/liquid hydrogen fuel cells to generate electricity and produce water as a byproduct. Based on the "Foton" fuel cells built for Buran, they are stacked in three sets, each capable of producing 2.5kWt of electricity and providing triple redundancy. The fuel cells will use the same liquid oxygen supplies stored on board for the engines. Liquid hydrogen supplies will amount to 7kg. With an overall mass of about 200kg, the fuel cells are significantly lighter than the solar panels that up until 2004 were planned to be installed on the Aggregate Compartment.

In the lifting-body configuration, the fuselage is equipped with two rudders and two body flaps and would have a cross-range capability of up to 500km (10 times more than Soyuz). Kliper would descend on parachutes, stowed in a container in the top section of the crew cabin. The deployment sequence would begin with a pilot chute pulling out a drag chute, which then in turn would extract the three canopies of the main chute. Kliper would be able to make a safe landing on a single canopy if needed. Alternative proposals using a back-up and main chute system (as on Soyuz) or aerodynamically shaped parachutes were rejected.

In order to ensure its reusability, the Return Vehicle needs additional systems to lower its landing speed and to prevent its direct contact with the ground. Ideas to use a ski landing gear or airbags were turned down in favour of a system of pneumatic shock-absorbers and solid-fuel rocket motors. Two pneumatic shock-absorbers located on the underside of the fuselage would be deployed at an altitude of 3.5km. With just 1.5m to 2m to go to touchdown, a gamma-ray altimeter would then command a 0.15sec burn of 23 soft-landing engines, 13 in the mid-lower section and 10 in the aft section of the fuselage to reduce vertical and horizontal speed respectively. These would ensure that the re-entry vehicle makes no unexpected turns as the shock-absorbers hit the ground. The parachutes are jettisoned immediately after touchdown.

The winged version of the fuselage is being studied jointly by RKK Energiya and the Sukhoi aviation design bureau (rather than NPO Molniya, Energiya's former partner in the Buran programme). It would make a classical horizontal runway landing using an aircraft type landing gear. This design would increase cross-range capability to 1,500/2,000km and lower the deceleration forces for the crew during re-entry. Moreover, with the absence of parachutes, shock-absorbers and solid-fuel engines, all of which are expendable systems, the degree of reusability would be higher. Runway landings also obviate the need for costly recovery operations in distant landing zones. Among the disadvantages of the winged design are its slightly heavier mass, its inability to perform ballistic re-entries and the lower chances of recovering the vehicle intact in case of a launch abort. In the latter event, reusablity of the cabin would be contingent on the vehicle's ability to reach a nearby runway.

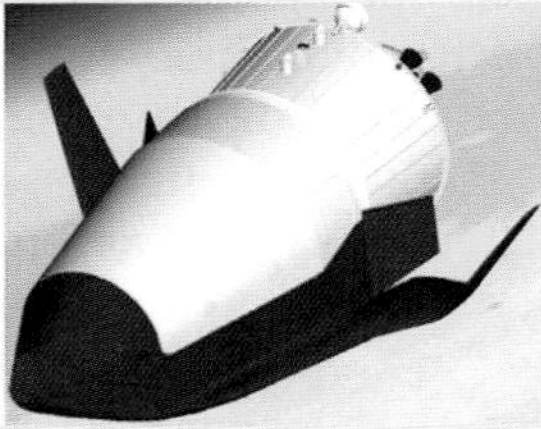

With the debate between lifting body versus winged design for Kliper, history repeats itself. The Russians faced a similar choice in the mid-1970s when they started working on a response to the US Space Shuttle. At that time plans for giant lifting bodies were abandonded in favour of the delta-wing Buran, essentially an aerody-

■ ***Above:*** *Two versions of Kliper : (left) a lifting body and (right) a winged vehicle. Images courtesy RKK Energiya.*

namic copy of the Shuttle Orbiter. For Kliper's design is the choice *not* to be driven by the US programme, which is aiming for a capsule type design with the Crew Exploration Vehicle.

It should be pointed out that the Russians have a certain flexibility in the choice of the fuselage since it is not an integral part of the crew cabin, which is compatible with either design of the fuselage. This means that they could opt to develop the lifting body first and introduce the winged design at a later stage.

Aggregate Compartment

The Aggregate Compartment's habitation compartment, situated behind the crew cabin, looks almost identical to the Soyuz orbital module and performs the same functions. Providing 8m³ of living space, it will among other things have a personal hygiene facility, sleeping bags and carry various consumables and life support systems. Mounted on the aft end will be an active Soyuz-TM type docking port, a new Russian-built Kurs-N rendezvous system (the current Soyuz version being built in the Ukraine), eight 24kg thrust attitude control thrusters and eight 50kg thrust manoeuvring engines. There are four clusters of engines, each containing two attitude control and two manoeuvring engines.

Surrounding the habitation compartment is a torus-shaped AO body which among other things will carry a thermal control radiator and propellant tanks for the aft engines and thrusters. Just as for the attitude control engines in the VA, the propellants are liquid oxygen and ethanol. Kliper will carry about 1,350kg of liquid oxygen and 750kg of ethanol. Tests of the engines, being developed in-house at RKK Energiya, were expected to get underway in late 2005.

The Aggregate Compartment is jettisoned from the rest of Kliper after the de-orbit burn and burns up on re-entry. The designers opted to include it as an expendable section to avoid problems associated with returning engines and tanks with propellant remnants and also to simplify the design of the parachute and landing systems [2].

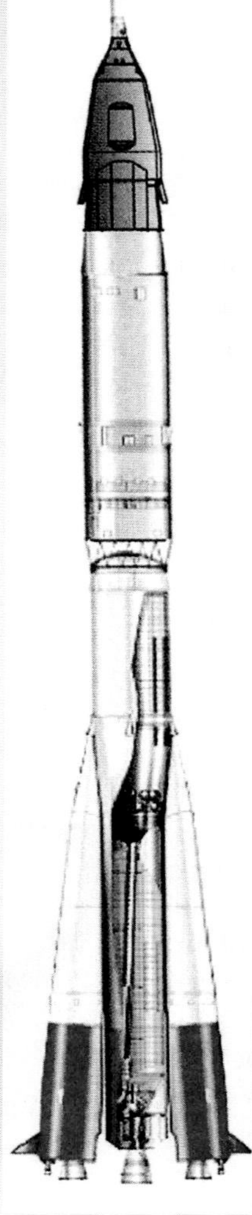

Centre: *Kliper mounted atop the Onega launch vehicle. Image courtesy Novosti Kosmonavtiki)*

Kliper and Parom

In the autumn of 2005 plans emerged for a lighter version of Kliper without the Aggregate Compartment. In this configuration Kliper would be launched into a low orbit by a lighter booster and link up with a space tug called Parom, which would tow it to the ISS. Although Parom has been under study at RKK Energiya since at least 2001, its existence was not revealed until April 2004. Parom is a 6.8 tonne vehicle that can be launched by a standard Soyuz rocket and would be stationed in orbit for five to ten years. Its initially announced mission was to pick up individually launched cargo containers in low orbit and then tow them to the ISS. Once a container is unloaded and stuffed full with waste, the tug returns it to a lower orbit, releases it and awaits the launch of the next one. Some of the containers would also have to carry propellant to refuel Parom.

Parom could perform the same task in combination with a stripped-down version of Kliper. In the absence of the Aggregate Compartment, the vehicle would need some other type of propulsion system to deorbit itself in case it fails to dock with Parom or in a nominal end of mission. That may be done with the help of a set of solid-fuel motors in an adapter attached to the aft end of the vehicle that could also be used to pull the vehicle away from the rocket in a launch abort. However, the Kliper/Parom plans remain vague at the time of writing. Possibly, the stripped-down Kliper is seen as an interim solution until a heavier rocket becomes availabe to launch the full-fledged version of the spacecraft [3].

The launch vehicle

Onega

The launch vehicle originally considered for Kliper was Onega. Named after a lake not far from St. Petersburg, it maintains the basic lay-out of the Soyuz rocket, but would be capable of lifting a much heavier payload thanks to the use of a larger propellant load and improved engines. Onega has its roots in two independent upgrade programmes for the Soyuz rocket. One of these is a government-funded project called "Rus", the goal of which is to gradually uprate the capacity of the rocket to

just over 8 tonnes by flying more capable engines and a modernized control system (Soyuz-FG, Soyuz-2-1A and 1B). "Rus" is managed by the Central Specialized Design Bureau (TsSKB) in Samara (formerly Kuibyshev). This bureau originated as Branch Nr. 3 of Sergei Korolyov's OKB-1 in 1959 to oversee the further development of R-7 based rockets and later also to design the nation's photoreconnaissance satellites. Aligned with the design bureau was a factory known as "Progress", where the bureau's rockets and spy satellites were integrated. Not eager to take part in the development of the Energiya rocket, Branch Nr. 3 became independent as TsSKB in 1974 and eventually merged with the Progress factory in 1996 to form TsSKB/Progress. Meanwhile, NPO Energiya set up a new branch in Kuibyshev (the so-called "Volga" branch) to design Energiya's giant core stage.

It was the Volga branch that embarked in the late 1980s/early 1990s on a separate effort to turn Soyuz into a medium-lift rocket capable of lifting over 10 tonnes. The Volga designers proposed a rocket with standard-sized strap-on boosters, but with a widened core stage powered by a single NK-33 engine, originally developed for the N-1 Moon rocket by the Kuznetsov design bureau, another Samara-based organization. The lower part of the core stage was widened from 2.06m to 2.66m and the upper part from 2.66m to 3.44m. When topped up to maximum capacity, the core stage would now carry 40-50 tonnes more fuel than the standard Soyuz core stage. The third stage was also widened to 3.44m to remain compatible with the core stage and because of that carried about 5 tonnes more propellant than the standard third stage.The rocket also borrowed two elements from the Rus programme, namely RD-107A (14D22) engines with improved fuel injectors in the first stage strap-on boosters (first flown on Soyuz-FG) and the new RD-0124E engine for the third stage (a modified version of the RD-0124 for the third stage of the Soyuz-2-1B). All these modifications would have enabled the rocket to deliver a payload of up to 11-12 tonnes into low Earth orbit. The rocket could also be adapted for missions to higher orbits by adding a fourth stage ("Taimyr" or "Korvet") inheriting technology from the Blok-L and Blok-D. Having virtually the same length as the standard Soyuz rocket, the new rocket required only minor modifications to existing Soyuz launch pads.

In 1996 the Volga branch proposed to use the four-stage version to launch RKK Energiya's Yamal communications satellites into geostationary orbit one at a time from Baikonur (rather than launching them in pairs on Proton rockets). RKK Energiya's central bureau in Moscow liked the idea and – in keeping with Russian tradition- named the rocket after its (expected) first payload. The three-stage version of the Yamal rocket was also eyed to launch Russian modules to the ISS, replacing the Ukrainian-built Zenit. However, there were no immediate prospects of receiving government funding for the project, which would have to be financed through private investments and loans. In 2000, RKK Energiya entered negotiations with the Asia Pacific Space

■ ***Above:*** *Kliper as it would look atop the Zenit launch vehicle. At left is the crew access tower built in the 1980s. Image courtesy RKK Energiya.*

Centre to launch a commercial version of Yamal called "Avrora" (Aurora) from Christmas Island in the Pacific, but financial problems eventually grounded the project.

In 2001, RKK Energiya began working on a rocket called Onega with the same dimensions as Yamal/Avrora, but specifically intended to launch a wide variety of Russian payloads from the Plesetsk cosmodrome, thereby minimizing Russia's reliance on Baikonur in Kazakhstan. It could launch Kliper into 62.8° and 64.8° inclination orbits from Plesetsk and into 51.6° and 64.8° inclination orbits from Baikonur. Although initial plans called for Onega's core stage to retain the NK-33, the ultimate plan was to equip it with the RD-191, developed by NPO Energomash for the core stage of the Angara rocket family. The strap-ons were to employ the RD-120.10F, a slightly modified version of the RD-120 LOX/kerosene engines used on the second stage of the Zenit rocket. The engine considered for the third stage was the RD-140E, a liquid oxygen/liquid hydrogen engine developed by the Design Bureau of Chemical Automatics (KBKhA) in Voronezh (also planned to be used on upper stages for Proton and Angara). For geostationary missions Onega would have a "Yastreb" upper stage using the RD-0126E, another liquid oxygen/liquid hydrogen engine developed by KBKhA [4].

Zenit

During a tour of RKK Energiya facilities on 30 November 2004, journalists were told that the prime contender for launching Kliper was the Zenit rocket, with Onega only considered as a back-up option. The reasons for not making this choice earlier must primarily have been of a political nature. First, despite the fact that Zenit uses engines developed by Russia's NPO Energomash, it is a Ukrainian launch vehicle, built by the GKB Yuzhnoe design bureau in Dnepropetrovsk. Second, it can only be launched from the Baikonur spaceport in Kazakhstan. Construction of a Zenit pad at Plesetsk *did* get underway in 1986, but it is now being rebuilt to support Angara launches. To make matters worse, there is only one single Zenit pad at Baikonur, the other one having been destroyed in a pad explosion in October 1990. Any bad accident on the remaining pad would therefore ground Kliper indefinitely. On the other hand, Zenit *was* originally intended to become a man-rated rocket. It was supposed to launch a Soyuz successor called "Zarya" studied by Energiya at the end of the 1980s and was also considered for launching small spaceplanes. One relic of those early plans is a crew access tower at the surviving Zenit pad, which could be refurbished for future Kliper launches.

A proposal to use Zenit for launching Kliper had come from Ukrainian President Leonid Kuchma in mid-September 2004 when the Presidents of Russia, Ukraine, Kazakhstan and Belarus met in the Kazakh capital Astana to move their countries closer to the ultimate goal of creating a free trade zone between them in a so-called Single Economic Space (SES).

The Kliper/Zenit proposal was to be approved at another summit of the four countries' presidents in St-Petersburg at the end of the year, but that was delayed indefinitely due to the political turmoil surrounding the Ukrainian presidental elections in late December 2004. It remains to be seen what effect the election of Viktor Yushchenko will have on the Kliper/Zenit proposal and future Russian-Ukrainian space cooperation in general, certainly in light of the fact that former President Leonid Kuchma once headed the plant where the Zenits are manufactured.

In August 2005, Russian Space Agency deputy head Nikolai Moiseyev announced that Zenit was still being considered as Kliper's launch vehicle. Kliper could benefit from the modernization of the Zenit launch infrastructure at Baikonur now being conducted in the framework of the "Land Launch" project. This is a commercial venture to launch satellites from Baikonur using the improved Zenit-M rocket, which essentially has the same modifications as the Sea Launch version of Zenit (Zenit-3SL). It would come in two versions, a two-stage version known as Zenit-M2 or Zenit-2SLB ("Space Launch from Baikonur") and a three-stage version called Zenit-M3 or Zenit-3SLB. Zenit-2SLB can launch 13.0 tonnes into low Earth orbit. Stanislav Konyukhov, the head of GKB Yuzhnoe, said later in the year that financial questions remain to be resolved, suggesting that Russia, Ukraine and Kazakhstan each contribute to the cost of the project [5].

Soyuz-2-3 and Soyuz-3

At the same time, it would appear that the Russians are again seriously considering Soyuz-derived launch vehicles to launch Kliper. Medium-lift Soyuz rockets were included in Russia's Federal Space Programme for 2006-2015 under the name "Rus-M", meaning that they can count on at least some government funding. The initial version, capable of orbiting 11 tonnes, is to be developed in 2007-2010 and a more advanced version with a payload capacity of 15 tonnes should appear in the 2010-2015 timeframe.

The initial version may be what some reports have referred to as the Soyuz-2-3 launch vehicle. All that has been revealed about it is that it can lift 11 tonnes by using an NK-33 engine and an increased propellant load in the core stage. In fact, it may be virtually identical to the earlier proposed Yamal and Avrora rockets.

The more powerful version could be a rocket known as Soyuz-3, first presented during Russian-European space talks in Moscow on 10 June 2005. Soyuz-3 appears to be a crossing between Yamal/Avrora and Onega, carrying a single NK-33 engine in the core stage, RD-120.10F engines in the strap-ons and four RD-0146E cryogenic engines in the third stage. This would enable it to launch a 14.0 tonne version of Kliper. A model of the Soyuz-3/Kliper combination was on display at the MAKS air show near Moscow in August 2005.

Possibly, Soyuz-2-3 will be used to launch the lighter version of Kliper that will be towed to ISS by the Parom space tug, while Soyuz-3 will orbit the full-fledged version of Kliper that can reach the station autonomously.

After its initial scepticism, TsSKB/Progress in Samara seems to have warmed to the idea of developing medium-lift Soyuz rockets, probably because Rus-M is a government-funded effort, unlike Yamal/Avrora/Onega. Moreover, there was a management change at the company in late 2003, when Dmitriy Kozlov retired as general designer after having headed the bureau for 45 years. At least one reason for the renewed interest in these boosters must have been the final go-ahead for the construction of a Soyuz launch pad at Kourou in French Guyana. This opens the prospect of launching Kliper from Kourou, making the project more attractive to potential European partners.

■ ***Above:** A model of Soyuz-3/Kliper on display at the MAKS air show in August 2005. Image courtesy Novosti Kosmonavtiki.*

Launch escape system

Kliper will be mounted on top of its launch vehicle, unprotected by a nose fairing. In the Onega configuration, Kliper had an emergency escape tower installed on its nose section. Outwardly very similar to the Soyuz emergency escape system, it would pull the Return Vehicle away from the rocket in an abort on the launch pad or in the early stages of the launch. However, later studies showed that in a pad abort a Soyuz-type emergency escape system would not be powerful enough to send the Kliper Return Vehicle to an altitude high enough for its parachutes to safely

deploy. An additional problem was that the escape tower would have to be attached to the vehicle's vulnerable nosecap, which is exposed to very high temperatures during re-entry.

Eventually, engineers came up with a totally different emergency escape system for the Zenit/Kliper combination. It consists of eight solid rocket motors installed on an adapter between the launch vehicle and the spacecraft. If no abort is required, the motors will still be used for the final orbit insertion manoeuvre, thereby providing extra weight savings. A similar escape system (weighing 3.8 tonnes) could be seen on the model of Soyuz-3/Kliper at the MAKS air show.

Status

Right now Kliper is still in the earliest stage of the design process ("Technical Proposals"), with work on the so-called "Draft Plan" (preliminary design) expected to be finished in 2007. So far RKK Energiya has financed Kliper development with its own means, but government funding could commence soon now that a new manned reusable space transportation system to succeed Soyuz has been included in the Federal Space Programme (FSP) for 2006-2015, officially approved by Russian Prime Minister Mikhail Fradkov on 22 October 2005. The Russian Space Agency expects to receive 10bn rubles (about €310m) in government funding for the new vehicle during that nine-year timeframe. Requirements are for the spacecraft to be at least 80% reusable and fly at at least 20 missions, carry up to six people and have a cargo capacity of at least 500kg up and down. The maiden flight should take place in 2013.

On 7 December 2005, Russian Space Agency head Anatoliy Perminov announced a closed tender would be held among three Russian space companies to build the new transportation system. Although Perminov declined to disclose who will be competing with RKK Energiya's Kliper, it was later reported that the two other contenders are NPO Molniya and the Khrunichev Centre. NPO Molniya is again

Above: *More views of Kliper. Images courtesy Nicolas Pillet.*

proposing its air-launched MAKS spaceplane, originally conceived in the late 1980s. The Khrunichev Centre is believed to have put forward a six-man spacecraft to be orbited by one of its Angara boosters and resembling an enlarged version of the return capsules developed for the Transport Supply Ships (TKS) in the 1970s. However, the requirements stipulated in the FSP are so obviously tailored to Kliper that one may wonder if this tender is no more than a formality.

There is no guarantee that Russian government funding will be sufficient to develop Kliper in a reasonable amount of time. In June 2006, ESA's ruling council agreed to invest some €15m in a two-year joint Russian-European study on the development of an Advanced Crew Transportation System capable of flying in low Earth orbit and to the Moon. This would be a capsule-type vehicle consisting of three compartments, casting further doubt on Kliper's future. A final decision on moving ahead with this programme is to be made by ESA ministers in 2008(6).

Missions

If Kliper is built in the forseeable future, it could play an important role in ISS operations. Officials have indicated it could be used to carry crew and cargo and also act as a lifeboat, with a single vehicle making it possible to evacuate a complete six-man resident ISS crew. Kliper could stay docked to the station for an entire year, twice as long as Soyuz and haul 500kg of cargo to the station and back to Earth.

The Russians also have other plans in mind for Kliper. They have said it could perform autonomous missions for "research purposes" or with tourists on board. In 2005, the Russian Space Agency was planning to carry out a study of the role that Kliper could play in ISS and Russia's "future orbital infrastructure". That orbital infrastructure may include a space station in a high-inclination orbit that would be used for Earth observations, applied research and support of future manned deep space missions. Kliper could also be adapted for missions to the Moon, linking up with a separately launched lunar lander and upper stage in the same fashion as America's Crew Exploration Vehicle. Kliper look-alike craft have even appeared as Earth return vehicles in RKK Energiya concepts for future manned Mars ships.

References :

1. V. Minenko: *"Recoverable Capsules and Transport/Research Space Vehicles"*, paper presented at a Mir symposium at the Dulles Hyatt Hotel in Washington on 27-28 July 1993 and published in "Mir-1 Space Station: A Technical Overview", NPO Energia LTD, Alexandria, 1994 ; Website of the Russian Agency for Patents and Trademarks.

2. S. Shamsutdinov: Project Kliper (in Russian), Novosti Kosmonavtiki, 7/2005, pp. 1-7.

3. R. Coppinger: Lighter Clipper could make towed trip to ISS, Flight International, 1 November 2005 ; T. Malik, *Russia's Next Spaceship : Alternative to NASA's CEV*, article on space.com website, 7 December 2005.

4. I. Chornyi: The far from weak Yamal (in Russian), Novosti Kosmonavtiki, 15-16/1998, pp. 49-51; *ISS Program – Progress M50 Mission*, Russian Space Agency press kit, 2004, pp. 36-40.

5. I. Afanasyev: About Land Launch (in Russian), Novosti Kosmonavtiki, 10/2005, pp. 66-67.

6. R. Coppinger: ESA boosts science, delays Kliper, Flight International, 13 December 2005

***Bart Hendricx** is a teacher of Russian and specializes in Soviet and Russian space history.*

www.springer.com/space www.praxis-publishing.co.uk

Other SPRINGER/PRAXIS titles on SPACE EXPLORATION

The Continuing Story of the ISS
Bond
PB 1-85233-567

Russian Spacesuits
Abramov/Skoog
PB 1-85233-732

Lunar Exploration
Ulivi/Harland
PB 1-85233-746

Space Shuttle Columbia
Evans
PB 0-387-21517

www.springer.com/space www.praxis-publishing.co.uk

The Sky at EINSTEIN'S Feet
Springer
PRAXIS
PHILIP CHIEN
COLUMBIA
FINAL VOYAGE
The Last Flight of NASA's First Space Shuttle

US Spacesuits
Kenneth S. Thomas
and Harold J. McMann
Springer
Apollo
The Definitive Sourcebook
Richard W. Orloff and David M. Harland
Springer
PRAXIS

Space Tourism
Van Pelt
CB&J 0-387-40213

Exploring the Moon
Harland
PB 1-85233-099

Life in the Solar System and Beyond
Jones
PB 1-85233-101

The Japanese and Indian Space Programmes
Harvey
PPC 1-85233-199

www.springer.com/space www.praxis-publishing.co.uk

Challenges of Human Space Exploration
Freeman
PB 1-85233-201

Creating the ISS
Harland/Catchpole
PB 1-85233-202

Russia in Space
Harvey
PB 1-85233-203

Neptune
Miner/Wessen
PB 1-85233-216

www.springer.com/space   www.praxis-publishing.co.uk

Other SPRINGER/PRAXIS titles on SPACE EXPLORATION

Project Mercury
Catchpole
PPC 1-85233-406

China's Space Program
Harvey
PB 1-85233-566

Mars - A Warmer, Wetter Planet
Kargel
PB 1-85233-568

Apollo - The Lost & Forgotten Missions
Shayler
PB 1-85233-575

www.springer.com/space www.praxis-publishing.co.uk

Other SPRINGER/PRAXIS titles on SPACE EXPLORATION

Mission to Saturn
Harland
PB 1-85233-656

Soyuz
Hall/Shayler
PB 1-85233-657

Walking In Space
Shayler
PB 1-85233-710

The Big Bang
Harland
PB 1-85233-713

www.springer.com/space   www.praxis-publishing.co.uk

THE END

The following pages: This spectacular 'blue marble' image is the most detailed true-colour image of the entire Earth to date. Using a collection of satellite-based observations, scientists and visualizers stitched together months of observations of the land surface, oceans, sea ice, and clouds into a seamless, true-colour mosaic of our planet. Image courtesy NASA Goddard Space Flight Center. Image by Reto Stöckli (land surface, shallow water, clouds). Enhancements by Robert Simmon (ocean colour, compositing, 3D globes, animation). Data and technical support: MODIS Land Group; MODIS Science Data Support Team; MODIS Atmosphere Group; MODIS Ocean Group Additional data: USGS EROS Data Center (topography); USGS Terrestrial Remote Sensing Flagstaff Field Center (Antarctica); Defense Meteorological Satellite Program (city lights).